# Go Green for Environmental Sustainability

# Go Green for Environmental Sustainability

## An Interdisciplinary Exploration of Theory and Applications

Edited by

Xavier Savarimuthu, SJ
Usha Rao
Mark F. Reynolds

CRC Press
Taylor & Francis Group
Boca Raton London New York

CRC Press is an imprint of the
Taylor & Francis Group, an **Informa** business

First edition published 2022
by CRC Press
6000 Broken Sound Parkway NW, Suite 300, Boca Raton, FL 33487-2742

and by CRC Press
2 Park Square, Milton Park, Abingdon, Oxon, OX14 4RN

---

**Library of Congress Cataloging-in-Publication Data**

---

Names: Xavier, S. (Savarimuthu), 1970- editor. | Rao, Usha (Geologist), editor. | Reynolds, Mark (Mark F.), editor.
Title: Go green for environmental sustainability : an interdisciplinary exploration of theory and applications / edited by Xavier Savarimuthu, Usha Rao, and Mark Reynolds.
Description: First edition. | Boca Raton : CRC Press, 2021. | Includes bibliographical references and index.
Identifiers: LCCN 2020056196 (print) | LCCN 2020056197 (ebook) | ISBN 9780367517403 (hbk) | ISBN 9781003055020 (ebk)
Subjects: LCSH: Industries--Environmental aspects. | Environmental protection. | Sustainable engineering.
Classification: LCC TD194 .G63 2021 (print) | LCC TD194 (ebook) | DDC 628--dc23
LC record available at https://lccn.loc.gov/2020056196
LC ebook record available at https://lccn.loc.gov/2020056197

---

ISBN: 978-0-367-51740-3 (hbk)
ISBN: 978-0-367-51741-0 (pbk)
ISBN: 978-1-003-05502-0 (ebk)

Typeset in Times
by KnowledgeWorks Global Ltd.

# Dedication

*I would like to dedicate this volume to the
sustainability warriors who inspire us all:*

*His Holiness Pope Francis, for the Encyclical
"Caring for our Common Home"*

*Rev. Fr. Arturo Sosa, SJ, the Superior General of the
Jesuits, for prioritizing the caring for our common home
as a Universal Apostolic Preference of the Jesuits*

*Mr. Rajinder Singh, an oasis of hope in the
water-starved state of Rajasthan, India*

*Vice President Al Gore, for his clarion call
on action for climate change*

*Mrs. Wangarĩ Muta Maathai, for her
environmental conservation efforts*

**Xavier Savarimuthu, SJ**

*We dedicate this book to:*

*Our wonderful son Luke who, like young people everywhere, deserves
to inherit a healthy planet from those who have gone before him,
and the memory of our beloved daughter Nina, whose light continues
to inspire us every day to do our part in building a better world.*

**Usha Rao and Mark F. Reynolds**

# Contents

## SECTION I   Agriculture, Bioremediation, and Green Chemistry

## SECTION II   Green Business, Banking, and Technology

# SECTION III   Environmental Sustainability

# SECTION IV   Green Engineering

# SECTION V  Green Healthcare

# SECTION VI  Green Philosophy

# Foreword

Humanity is faced with an urgent, complex, and vexing environmental crisis of a scale that is unprecedented in human history. The role of higher education in addressing this crisis is to assemble scholars from all disciplines to disclose the underlying social and economic drivers of environmental degradation and climate change, and develop a way forward that integrates human well-being with science and technology solutions.

This volume entitled, *Go Green for Environmental Sustainability: An Interdisciplinary Exploration of Theory and Applications*, is an excellent example of science and technological solutions to environmental problems being advanced by thoughtful scholars from multiple disciplinary and cultural perspectives.

The topics covered within are diverse, ranging from green chemistry and bioremediation to green business and technology, each chapter shining a light on new and innovative research in these areas. The common thread throughout the text is the need for urgent action, a vision for a sustainable future, and a mindfulness that solutions must be widely accessible and advance the welfare of all nations.

This compilation is an idea whose time has come. I am confident that this excellent collection will inform policymakers and industry leaders with ways to advance environmental sustainability in their cities, states, nations, and organizations. It will inspire new innovative academic research and experiential learning to prepare the next generation of environmental leaders from the higher education sector.

Compelled by the realization that we do not inherit this Earth from our ancestors but instead borrow it from our children, the ideas in this volume offer practical solutions, comprising an important contribution to the total worldwide effort to build a sustainable tomorrow.

**Nancy C. Tuchman, PhD**
*Founding Dean*
*School of Environmental Sustainability*
*Loyola University*
*Chicago, Illinois*

# Preface

We write this as the world reels from the devastation unleashed by the COVID-19 pandemic. Mahatma Gandhi presciently warned us, "The world has enough for everyone's needs, but not everyone's greed." This pandemic has laid bare humanity's fathomless greed that destroys the very environment that is essential to human survival. In addition to its many grim effects, the pandemic has also served to turn the world's eye away from the far more dangerous reality of accelerating global climate change and the urgent need for sustainable development.

The term Sustainable Development was introduced in the United Nations Brundtland Commission Report, Our Common Future, produced in 1987 under the leadership of Gro Brundtland, the former Prime Minister of Norway and Chair of the World Commission on Environment and Development. Although there are many definitions of sustainable development, it is commonly described as the ability to meet the needs of today without compromising the ability of future generations to meet their own needs.

We should understand sustainable development as a careful reassessment of a balance sheet where the debits (read: exploitation of natural resources) and credits (read: socioeconomic human development) are set off against each other to arrive at the future that we leave behind for children everywhere. The world's major religions stress the importance of sustainable living as an urgent need for those of any faith and no faith. Global leaders such as Pope Francis and the Dalai Lama have emphasized the urgency of sustainable development and the need for caring for our common home, so that we do not compromise the ability of future generations to live, work, and thrive on this planet. The Jesuits, an order to which one of us belongs, in their Universal Apostolic Preferences also emphasize the need to care for the entire planet as an integral part of the Jesuit mission for the next decade.

This book is therefore an effort to broker a peace between economic growth and sustainable climate action, before the onset of a new geologic era of our own making leaves no room for humans.

**Xavier Savarimuthu, SJ**
**Usha Rao**
**Mark F. Reynolds**

# Acknowledgments

The editors would like to acknowledge the many individuals without whose support the dream of this book project could not have been realized. We would like to express our gratitude to Ms. Cindy Renee Carelli and Ms. Irma Britton, Executive Editors, and Ms. Erin Harris, Senior Editorial Assistant, CRC Press, Taylor & Francis Group, for their able guidance and ready support through all stages of this project, especially since publishing a book during a raging global pandemic brought a plethora of unforeseen challenges. We are grateful to Dr. Tapalina Bhattasali for her able and generous support in serving as a liaison between CRC Press and the editors. Ms. Deirdre Taft-Lockard lent her excellent editing and indexing skills to the project. Mr. Suvobrata Ganguly, Editor, Core Sector Communique, is thanked for providing the cover photo. Dr. Nancy Tuchman, Founding Dean of the School of Sustainability, Loyola University, Chicago, is gratefully acknowledged for her thoughtful exposition in the foreword that gives credence to the book's objective. We also offer heartfelt appreciation to all the contributors for their work on this project.

Xavier Savarimuthu, would like to thank the Almighty for the profound generosity that He has bestowed upon us, even during such challenging times as these. The Society of Jesus is gratefully acknowledged for constantly providing support of the author's research. St. Xavier's College (Autonomous), Kolkata, also played a pivotal role by providing the necessary infrastructure. X.S. would also like to thank his co-editors Usha Rao and Mark F. Reynolds for their ready participation in this project, notwithstanding their many research and teaching commitments. Grateful acknowledgment is due to Rev. Fr. Arul Sivan, Rev. Fr. Anthony Joseph, and Fr. John Pradeep at Mount St Joseph Jesuit Community, Bengaluru, for mentoring during the publication process; to Rev. Dr. Sebasti L Raj, Rev. Dr. Xavier Vedam, Rev. Dr. Ignacimuthu, Rev. Dr. John Rose, Rev. Fr. Johnson Padiyara, Rev. Fr. Dejus J R, Rev. Dr. Joseph Varghese, for their unflinching support and encouragement throughout X.S.'s career; to Prof. Allan H. Smith, University of California, Berkeley, for inspiration during the formative years as a budding environmental scientist; and to Prof. Protap Chakravarti, Former Director, Geological Survey of India, and an internationally renowned geoscientist, for contributing to his development as a scientist and researcher.

Usha Rao and Mark F. Reynolds wish to acknowledge Fr. Xavier Savarimuthu for conceiving of this project and working with us to bring it to successful completion. We want to thank our families for their support and inspiration throughout our careers. Usha gratefully acknowledges her parents Mr. Raja Gopala Rao and Mrs. Purna Kumari Rao who are steadfast sources of support through good times and bad, and excellent role models for always prioritizing the common good over narrow interests. We also want to acknowledge many current and former students of Saint Joseph's University who have assisted us in our respective research programs and taken our courses over the decades. This book is as much theirs as it is ours.

# Editors

 **Xavier Savarimuthu, SJ, PhD,** was a Vice-Principal and Head of the Department of Environmental Studies at St. Xavier's College, Kolkata. He has extensively presented and published papers on his work on arsenic pollution both in India and abroad. His latest article entitled, "The Future We Want Requires Reconciliation with Creation," was published in four different languages, namely, French, Spanish, Latin, and English (https://www.luc.edu/media/lucedu/ijep/documents/PJ111ENG.pdf). He has delivered invited lectures in Stockholm, Sweden; Manila, the Philippines; Paris, France; Bonn, Germany; and at the University of Oxford. He has taught at Santa Clara University, California, and Saint Joseph's University, Philadelphia, where he held the endowed Donald I. MacLean Jesuit Chair for accomplished scholars and teachers. He has ignited the minds of many young people in the field of environmental science, creating a new generation of ecologically conscious citizens. He has co-authored a textbook for undergraduate and graduate science and engineering students entitled, *Fundamentals of Environmental Studies,* published by Cambridge University Press.

 **Usha Rao, PhD,** is an Associate Professor of Environmental Geochemistry at Saint Joseph's University, Philadelphia, Pennsylvania. She obtained her undergraduate degree in geology from the University of Bombay, and her doctoral degree in geochemistry from the University of Rochester, prior to completing Environmental Research Council funded post doctoral work at Northwestern University. At Saint Joseph's University where she has been an award-winning teacher for over twenty years, she is the Founding Director of the Office of Teaching and Learning, and the co-founder of a mentoring program for gifted STEM students. Her research on environmental pollution has been widely published and supported by grants from the American Chemical Society – Petroleum Research Fund, the Christian R. and Mary F. Lindback Foundation, and the National Science Foundation. She serves on the editorial board of the international journal *Environmental Geochemistry and Health*, published by Springer Nature. She has served as a Climate Reality Mentor for US Vice President Gore's Climate Reality foundation, and as an Approved Expert Reviewer for the UN Intergovernmental Panel on Climate Change (UN IPCC). She serves as a scientific consultant to corporations and intergovernmental organizations, and as a board trustee of several non-profit organizations. Dr. Rao was recognized by the Philadelphia Chapter of the Association for Women in Science with the 2021 Elizabeth Bingham Award, given to "distinguished scientists who have significantly influenced the advancement of women in science."

 **Mark F. Reynolds, PhD,** is an Associate Professor of Biochemistry at Saint Joseph's University in Philadelphia. He received his undergraduate degree in chemistry from Grinnell College as a Charles White Scholar, and his PhD in Inorganic Chemistry from the University of Wisconsin–Madison as a National Science Foundation Biophysics Trainee. He was a post-doctoral researcher at the University of Minnesota. At Saint Joseph's University, he currently serves as the Director of Saint Joseph's University's Office of Competitive Student Fellowships. His research focuses on the role of heme-based gas sensing proteins which sense NO, CO, and $O_2$, and play a crucial regulatory role in a wide variety of organisms as well as heme-responsive proteins that are affected by binding heme, fields in which he is a recognized pioneer. His research has been published in international scientific journals including *Nature*, the *Journal of Biological Inorganic Biochemistry*, and *Biochemistry*, and was profiled in the American Chemical Society's *Chemical and Engineering News* and in the American Association for the Advancement of Science's *Eureka Research Highlights*. Dr. Reynolds' research has been funded by the American Chemical Society – Petroleum Research Fund, Research Corporation, Merck, Inc., the National Institutes of Health, and the National Science Foundation.

# Introduction

## EXPLORING WHAT IT MEANS TO BE SUSTAINABLE THROUGH AN INTERDISCIPLINARY AND GLOBAL LENS: APPLICATIONS FOR A GREENER TOMORROW

The unsustainable impact of humankind on the Earth's environment is now laid bare by the accelerating and worldwide reverberations of global climate change. We are at a critical crossroads in human history, where the need to embrace new and creative solutions to stem the tide of environmental change has gained urgency. This research volume, *Go Green for Environmental Sustainability: An Interdisciplinary Exploration of Theory and Practice*, is conceived as a way to bring an interdisciplinary perspective to these problems. Although most of the manuscripts contained within originate in the STEM fields of biology, chemistry, environmental science, and structural, civil, environmental, and computer engineering, the issues are too thorny and multifarious to be solved within any one discipline or related set of disciplines. Thus, we have included expert voices from disciplines such as economics, business, and environmental philosophy in the collection. Another goal of the book has been to pull together emerging strands of science-based research from developed and developing countries in recognition of the fact that any ethical and lasting solutions must necessarily encompass all people, everywhere. As such, we offer articles that examine green practices and challenges in fields such as agriculture, microbial and enzyme bioremediation, and green chemistry, green business, technology and engineering, and sustainable healthcare, all examined across the global North-South equatorial divide, to name a few of these threads. It is only with a deliberate and concerted effort on the part of academic researchers and teachers, governmental and intergovernmental regulatory and oversight bodies, and industry practitioners, that we can hope to leave our small blue planet in good enough shape to sustain many more millennia of human existence.

In the first section of the book, chapters by Reynolds, Sultana, and their respective co-authors explore the promise and potential of bioremediation through synergistic microbial consortia and selective enzyme degradation, with a focus on removing toxic textile dyes and heavy metals such as cadmium, lead, and zinc from industrial effluent, a public health problem that plagues many countries that have large textile industries and unregulated discharge to surface drinking water and groundwater. Our authors also turn their lens on agriculture, which is beset by problems ranging from the need to feed ever-increasing numbers of humans using the same amount of arable land, to water shortages, soil loss, and pollutant discharges. The escalating impacts of climate change on agriculture is largely under-recognized by the general public, even though in recent years, climate change has reduced the viability of certain food crops even in technologically developed countries such as the United States. Chapters by Rao and Saul, and T. Sultana and co-authors critically assess emerging areas in biogeochemistry such as the use of beneficial soil microbial communities, semiochemicals, biopesticides and biofertilizers, nutrient uptake enhancers, and nanotechnology, which may be harnessed to solve the growing and severe problems in agricultural production. These newer developments have the additional advantage of reducing traditional pesticide and other agricultural chemical discharges that pollute land, water, and air, and form threats to both human and ecosystem health, in addition to reducing the greenhouse gas effects of agriculture, which is a major contributor to global climate change.

The science that provides an underpinning for a paradigm shift in sustainability can be connected to new technologies that help businesses understand environmental practices to be a fundamental part of their function. Schellhorn and Burkhalter argue that a single-minded pursuit of economic growth has left no room for environmental justice, and that financial markets must embrace both mitigating climate change risks and building inclusive prosperity as mutually beneficial. Roy and Savarimuthu demonstrate that public and private sector banks that adopt green banking initiatives, by no means a common practice in all parts of the world, reduce their costs and expenses, as compared to those that follow traditional banking practices. The following chapters explain how green computing and artificial intelligence can help lower our energy consumption and carbon footprint. One chapter from the discipline of computer engineering examines how the increasing demand for information technology, which is a significant consumer of electricity and total energy, can be mitigated by green practices such as e-waste minimization and targeted recycling, virtualization, cloud computing, and green data centers. The other chapter on this same theme explores how Industry 4.0 with its disruptive technologies such as artificial intelligence, machine learning, deep learning, sensor technology, Internet of Things, and other developments can find use in agriculture, water management, waste management, and remote, targeted environmental monitoring to achieve sustainable development goals.

These ideas are built on and extended in the section that follows, with Beed and co-workers examining how green transportation practices such as private vehicle sharing, currently only common in large urban centers in the highly developed countries, can find widespread use throughout the world, thus reducing the burden on public transportation networks and providing flexible transportation options. The next chapter focuses on many environmental measures undertaken by the Scandinavian countries which lead the world in green development. These initiatives in infrastructure, waste, and energy management, among others, can provide a tried-and-tested model for all nations to emulate. The study that follows offer a counter-example on the national scale, tying the Australian government's hesitancy to directly address climate change with the disastrous bushfires of 2019–2020 that destroyed an astounding 14 million acres of land, displaced over 90,000 people, and cost billions of dollars – all in a developed nation that could work to mitigate these and other climate risks. The next study by Badagha and co-workers in structural and civil engineering explains how industrial waste can be used in place of cement to make new types of concrete that are environmentally friendly, in search of a circular economy and waste-free society. Cement and concrete production are major contributors to greenhouse gas emissions, so this innovation packs a double whammy by minimizing climate-damaging cement manufacture, and finding a way to reuse and recycle industrial waste. The authors explore engineering material properties such as strength, durability, and others that increase the potential for waste material to be reused in this manner. Ghosh examines the disclosure of environmental issues by the mining sector using a constructed disclosure index and argues for better environmental reporting and disclosure practices, which will provide assurance to the public and other stakeholders that businesses will be accountable for any environmental destruction that they cause.

The enormous and expanding healthcare industry, synonymous with healing to many, in fact also contributes to environmental pollution in different ways, accounting for 10% of greenhouse gas emissions, 10% of smog formation, and nearly 10% of all air pollution in the United States alone. Ironically, the total number of deaths and disease from exposure to the cumulative effects of air, water, and land pollution are believed to equal the number of deaths and disabilities incurred by physician and hospital mistakes. However, while hospitals, medical training programs, governments, and patient safety advocacy organizations, all work tirelessly to limit the effects of the adverse effects of medical treatment and human error on patients, the deleterious effects of healthcare-created pollution on the general public have remained all but invisible. The next two studies described in our volume offer valuable insights into this emerging and nascent field of green healthcare. Mandal evaluates the marketing of greener pharmaceuticals to physicians using statistical methods such as factor analysis and cluster analysis to assess physicians' self-reported value orientation with respect to the environment, and concludes that a viable market exists in green pharmaceuticals. The next chapter explores how hospitals can embrace green building infrastructure using the principles of Building Information Modeling (BIM) and energy modeling strategies, and incorporate greener cleaning practices into hospital management. In the final work, philosopher Kuperus discusses the role of the individual as an integral part of the environment, not standing apart from it and above other animals, and draws on religious principles from the Eastern faiths to argue that individualism needs to make way for a more expansive worldview that includes community and nature in an effort to build a sustainable world and worldview.

The research profiled in this volume collectively demonstrates that the hope of achieving a sustainable society still lies within our reach, provided we undertake urgent and focused efforts in achieving inclusive environmental sustainability for all nations and all peoples, looking back only to learn from our mistakes and looking forward to the dream of a brighter tomorrow.

# Section I

*Agriculture, Bioremediation, and Green Chemistry*

# 1 Making Heme Proteins Green

## Designing Enzymes for Bioremediation and Environmental Sustainability

*Mark F. Reynolds, Daniel Barry, and Sarah Pogash*

## CONTENTS

## 1.1 INTRODUCTION

Modern industry and agriculture have had a profound impact on the environment and have led scientists to pursue new research into green chemistry and bioremediation toward an environmentally sustainable future. Industrial dyes and halophenols are one of the biggest causes of environmental contamination and water pollution in the world, and are one of the major contributors to industrial water pollution (Lellis et al. 2019). Over 90,000 tons of water are used annually in the textile industry and at least 20% of the industrial water pollution comes from textile dyes (Routoula and Patwardhan 2020). This is even more important in Asia, where dye use contributes to a significant portion of the water pollution. India is one of the largest textile manufacturers in South Asia, with 90% of the textile production occurring in the western states of Maharashtra and Gujarat (Gita et al. 2017). Some of these organic dyes are mutagens and carcinogens that can lead to human health problems and the deterioration of ecosystems (Lellis et al. 2019).

In Mumbai, Maharashtra, for example, the contamination of the Mithi River from industrial dumping and untreated water has been a recent source of lawsuits and controversy. The NGO Watchdog Foundation in 2017 identified the textile industry, the laundry detergent industry, the ready mix concrete and marble industries, the powder cutting industry, and vehicle service centers as the major contaminators of the Mithi River. A 2009 study published by the Human Health Report found that people living in Mumbai have an average life expectancy that is seven years less than the average person in India, and industrial water pollution is thought to play a key role in this disparity. Thus, techniques for removing toxic industrial dyes from water have gained increasing urgency in India and many other countries across the globe.

Industrial dyes can be classified into several categories, the most common being the azo dyes that make up approximately 60% of the dyes, and the next most common being the anthraquinone dyes that are even harder to degrade because of their high redox potentials (Routoula and Patwardhan 2020). These dyes are water-soluble organic compounds that persist in the environment. In contrast, halophenols are commonly used as pesticides or insecticides and can cause damage to the liver and immune systems (Godoy et al. 1999). However, both textile dyes and halophenols are considered carcinogens and potential mutagens, and thus there exists a real need for new methods that can remove them safely from water (Lin 2019).

Physical methods such as high temperature incineration and chemical degradation have been used to remove industrial dye pollutants and halophenols (Ghaly et al. 2013, Guo et al. 2013). However, more efficient and environmentally sustainable methods are needed (Routoula and Patwardhan 2020). Biodegradation has shown promise because the process is both economically and environmentally friendly. In fact, enzymes from a variety of organisms such as bacteria and fungi have been used to degrade pollutants in a wide range of bioremediation efforts (Lin 2019). Thus, scientists are looking for new and genetically engineered enzymes that can degrade textile dyes and other toxins found in industrially polluted water.

This chapter focuses on the role of heme-containing enzymes in bioremediation. It is not surprising that heme proteins have played an important role in biodegradation efforts. Heme proteins are one of the oldest families of proteins, and play important roles such as oxygen binding and delivery (myoglobin and hemoglobin), electron transfer (cytochromes), and catalysis (Cyt P450s and peroxidases), and are found in all kingdoms of life (Poulos 2014). Several families of heme proteins have been discovered in the past decade that can degrade industrial contaminants, including the dye-decolorizing peroxidases (DyPs) and the dehaloperoxidases (DHPs) that can degrade halophenols (Lin 2019). Also, well-studied heme proteins such as myoglobin and cytochrome $c$ peroxidases have been genetically engineered to serve as bioremediation catalysts (Lin 2019).

In addition, there are many new families of heme proteins that could be used in the bioremediation of toxic industrial products. Heme-based gas sensing proteins are ubiquitous in nature and can sense carbon monoxide (CO), oxygen ($O_2$), or nitric oxide (NO), and regulate a wide variety of important biological processes such as nitrogen fixation, blood pressure, and circadian rhythm (Gilles-Gonzalez and Gonzalez 2005, Shimizu et al. 2015). The biological diversity and abundance of heme-based gas sensing proteins make them a potential target for bioremediation efforts. This chapter explores the potential for using native and genetically engineered heme proteins to remove toxic textile dyes and halophenols from industrially polluted water.

## 1.2 DYE-DECOLORIZING PEROXIDASES

Heme peroxidases are a large superfamily of heme enzymes that utilize hydrogen peroxide to oxidize a range of substrates (Poulos 2014). More recently, a distinct family of heme peroxidases enzymes, the DyPs have been discovered in bacteria, fungi, and eukaryotes (Singh and Eltis 2015, Linde et al. 2015, Yoshida and Sugano 2015). These are able to oxidize a wide range of anthraquinone dyes like blue number 19 (RB 19) which are difficult to degrade, although the native substrates for most of these enzymes have not been identified (Singh and Eltis 2015, Linde et al. 2015, Yoshida and Sugano 2015). A variety of DyPs have shown promise in early bioremediation efforts; however, their catalytic rates are less than ideal, with catalytic rates significantly lower than heme peroxidases. Additionally, DHPs favor acidic pH. Therefore, improved bioremediation catalysts are needed (Lin 2019).

The first DyP-type enzyme from the fungus *Bjerkandera agusta* was identified in 1999 (Kim and Shoda 1999). It is generally accepted that there are at least four different classes of DyPs (A, B, C, and D) (Singh and Eltis 2015), although Yoshida and co-workers have argued that there are only three DyP classes: P (primitive), I (intermediate), and V (advanced). The authors believe that the classes C and D DyPs are similar in structure and phylogeny based on their analyses (Yoshida and Sugano 2015). DyPs from different sources and families differ in their catalytic and structural properties, although they have many similarities such as dye degradation (Singh and Eltis 2015, Linde et al. 2015, Yoshida and Sugano 2015). Work has focused on identifying the intermediates in the reaction cycle, and, similar to the heme peroxidase family, they can generate compound I and II reaction intermediates (Pfanzagl et al. 2018), although not all Dyps do. Also, while heme peroxidases are able to degrade the azo textile dyes, only DyPs are capable of degrading anthraquinone dyes. It is also clear that the DyPs are a distinct family of heme peroxidases based on the amino acid sequence and the initial proteins that have been crystallized and structurally characterized, although much work remains to be done (Singh and Eltis 2015, Linde et al. 2015, Yoshida and Sugano 2015).

Structurally, DyPs contain a C terminal two domain $\alpha + \beta$ ferredoxin-like fold that is distinct from the structures of the heme peroxidase superfamily of enzymes (Singh and Eltis 2015, Linde et al. 2015, Yoshida and Sugano 2015). Although there is diversity among the classes of DyPs, many of the amino acids in the heme binding active site are conserved. All of the DyPs contain a conserved proximal histidine ligand. On the distal side of the heme pocket, where catalysis occurs, there exists a bound water molecule in the resting, $Fe^{3+}$ state of the heme. In addition, there is a conserved aspartate and arginine pair on the distal side of the heme that has been proposed to play a key role in catalysis (Pfanzagl et al. 2018). The aspartate could act as a catalytic base by removing a proton from hydrogen peroxide, and the arginine would then help stabilize the resulting intermediate. This is based on the similarity to the heme peroxidase superfamily which uses a histidine base and an arginine to stabilize the charge. Site-directed mutagenesis studies of DyPs where the distal aspartate and arginine were removed showed that the aspartate is essential for catalysis while the arginine was only essential in some DyPs, although more work needs to be done on their detailed mechanisms (Pfanzagl et al. 2018). Tryptophan and tyrosine residues in the heme second coordination sphere were also shown to play an important role in catalysis. Also, a tryptophan on the surface of the protein was shown to be important in the oxidation of the dye blue 19 (Shrestha et al. 2016). Thus, DyPs have both an aspartate-arginine distal pair and a tryptophan-tyrosine pair that appear to play key roles in the peroxidase reactions of DyPs. The heme site of DyPs is clearly distinct from the rest of the heme peroxidase superfamily which allows it to oxidize a variety of difficult-to-degrade anthraquinone dyes found in textiles.

Because DyPs are hard to isolate and characterize, have an acidic pH optimum, and have relatively low catalytic activities compared to other heme peroxidases, researchers have looked into generating DyPs-like enzymes with increased catalytic activity by genetically engineering DyPs as well as other heme proteins. Lin and co-workers discovered that by introducing a phenylalanine and a tryptophan on the distal side of myoglobin using site-directed mutagenesis, they were able to approach, and in some cases exceed, the catalytic rates of native DyPs (Zhang et al. 2019). In the case of cytochrome $c$ peroxidase, researchers introduced several tryptophan amino acids from the heme catalytic site to the surface of the protein, making a tryptophan wire that allowed for the catalytic degradation of several dyes (Field et al. 2017). It is clear that new and improved enzyme catalysts are needed to to make them more feasible for bioremediation efforts.

## 1.3 DEHALOPEROXIDASES

A new class of heme peroxidase enzymes was discovered in the mid-1990s in the sea worm *Amphitrite ornate* that can survive under conditions with toxic halophenols produced by other worms (Chen et al. 1996). This new class of enzyme was named the DHPs because they can use hydrogen peroxide to catalyze the conversion of mono-, di-, or tri-halophenols to less toxic quinone products (Franzen and Thompson 2012). Dehaloperoxidases can degrade not only the natural bromophenol substrates, but also toxic industrial pollutants such as 2,4,6-trichlorophenol (TCP) (Franzen and Thompson 2012). Thus, the dehaloperoxidases are an important target for bioremediation efforts.

Over the past two decades, researchers have learned more about the catalytic activity and function of DHPs. There are two DHP isozymes, A and B, that differ by only 5 amino acids (I9L, R32K, Y34N, N81S, and S91G), out of 137 amino acids. It has been proposed that these two proteins may form an $\alpha\beta$ heterodimer, or an $\alpha_2\beta_2$ heterotetramer like hemoglobin, in solution (D'Antonio et al. 2010). The B enzyme is almost three times faster at degrading 2,4,6-tribromophenol and 2,4,6-trichlorophenol than the A enzyme, The B form of DHP also exhibits a tenfold greater catalytic efficiency ($K_{cat}/K_M$) compared to the A enzyme (D'Antonio et al. 2010).

Structurally, DHPs have a globin-like fold, with a proximal histidine bound to the heme iron, and a distal histidine that can hydrogen bond to the substrate (de Serrano et al. 2007). There are also several hydrophobic tyrosine and phenylalanine residues on the distal side of the heme that may play an important role in substrate binding and catalysis. A recent structural study of a Y34N DHP A variant with bound TCP revealed that there is an internal binding site in the heme pocket as well as an external binding site (Wang et al. 2013). There is also an important hydrogen bonding interaction between the tyrosine 38 and the hydroxyl group of the TCP substrate. The authors propose that TCP binds first to the heme site through hydrogen bonding interactions, then hydrogen peroxide binds to trigger the reaction (Wang et al. 2013). Because DHPs are not particularly fast enzyme catalysts, improvements need to be made in order to use them as efficient bioremediation catalysts.

Recently, several groups have looked at artificial DHPs and the use of artificial heme proteins such as myoglobin and horseradish peroxidase (HRP) to speed up the dehalogenation reaction (Lin 2019). Myoglobin has a similar globin fold to the DHPs as well as containing both the proximal and distal histidine residues, and the native form already has some peroxidase activity (Osborne et al. 2007). Researchers have replaced residues in the heme site of myoglobin to attempt to make a more effective catalyst (Lin 2019). The most effective myoglobin mutants designed so far contained the F43Y/H64D double mutant where a distal phenylalanine is replaced with a tyrosine that hydrogen bonds to the TCP substrate in the crystal structure, and the distal histidine has been changed to an aspartate which presumably helps with catalysis (Yin et al. 2018). A horse-radish peroxidase-like mutant, M86E DHP, was shown to have enhanced dehalogenation catalytic activity (Du et al. 2011). In addition, a *de novo* designed four helix bundle, called a C45 machete, with a *c*-type cytochrome that was changed to have an open site at the heme, has enhanced dehalogenation activity (Watkins 2017). Because this C45 machete can be produced *in vivo* from post-translational modification of cytochrome *c* and is very stable, it has great potential for green chemistry and bioremediation.

## 1.4   HEME-BASED GAS SENSING PROTEINS

Since the surprising discovery of heme-based gas sensing proteins in the early 1990s, the field has expanded dramatically to include proteins from archaea, bacteria, and eukaryotes (Gilles-Gonzalez and Gonzalez 2005, Shimizu et al. 2015). These heme proteins sense $O_2$, NO, or CO and control a wide range of important biological processes such as blood pressure regulation, circadian rhythm, chemotaxis, and nitrogen fixation (Gilles-Gonzalez and Gonzalez 2005, Shimizu et al. 2015). The heme-based gas sensors proteins have been placed into families based on their amino acid sequence and three dimensional structures. These families include the heme-NO binding (HNO-X) family, the CO-sensing CooA family, the heme-PAS family, the globin coupled receptors (GCS) and several other new families. (Shimizu et al. 2015). Heme-based gas sensing proteins typically contain an N-terminal heme domain that senses the $O_2$, NO or CO gas, and a C-terminal effector domain that transmits the biological signal (Gilles-Gonzalez and Gonzalez 2005, Shimizu et al. 2015).

Because of the variety of heme binding sites, the diversity of redox potentials and reactivity, and their ubiquitous nature, heme-based gas sensing proteins make an ideal target for bioremediation efforts. There is great demand for environmentally friendly catalysts that can remove toxic and difficult-to-degrade pollutants from water. Heme-based gas sensing proteins have evolved a variety of new heme coordination environments that might prove useful in biodegradation. This chapter probes the uses of native and genetically engineered heme proteins as environmentally friendly green catalysts to remove pollutants from contaminated water streams.

## 1.5  HEME-BASED NO SENSORS (H-NOX)

The heme-based NO sensors (H-NOX) were the first heme-based gas sensing family to be characterized when researchers discovered the role of NO in blood pressure regulation (Waldman and Murad 1987). In humans, the NO gas is formed in the endothelial layer of blood vessels by nitric oxide synthase and diffuses to the smooth muscle layer where it binds to the heme iron of soluble guanylate cyclase (sGC), the NO sensor, which activates the enzymatic conversion of guanosine triphosphate (GTP) to cyclic guanosine monophosphate (cGMP) (Waldman and Murad 1987, Wolin et al. 1982). The cGMP is the second messenger that triggers the phosphorylation and opening of calcium channels in the smooth muscles, which lowers blood pressure.

The mammalian NO-sensor sGC is a heterodimeric heme protein with an $\alpha$ subunit that contains the heme domain and a $\beta$ subunit that helps to form the catalytic heterodimer (Waldman and Murad 1987). Studies have shown that NO binds to the reduced, $Fe^{2+}$ five-coordinate heme, which triggers the loss of the proximal histidine from the heme iron, and causes the dramatic activation in the production of the second messenger cGMP (Reynolds and Burstyn 2001, Deinum et al. 1996, Yu et al. 1994). Although researchers have been unable to obtain a crystal structure of full-length mammalian sGC, truncated forms of sGC and bacterial H-NOX proteins have led to important structural studies and the characterization of this family.

The heme sites of the bacterial H-NOX proteins have been structurally characterized and shown to contain a five-coordinate heme with a proximal histidine bound to the heme iron and a distal tyrosine which plays an important role in the NO activation mechanism (Shimizu et al. 2015). There are several features to look at when determining whether a heme protein would be a good bioremediation catalyst. One of the most important features is the coordination chemistry at the heme iron and the second coordination sphere environment, particularly, on the distal side of the heme where substrates and hydrogen peroxide react. There is space for substrates to bind to the distal side of the heme in the crystal structures of bacterial NO-sensing proteins, but there is a distal tyrosine instead of the conserved distal arginine/aspartate pair in the DyPs and a distal histidine in the dehalogenation reactions of DHPs. The distal tyrosine of these H-NOX proteins could be replaced although it remains unclear how that replacement would affect the heme site.

In addition, the H-NOX proteins have a high redox potential that favors the reduced $Fe^{2+}$ state of the heme. In fact, the heme-based NO sensors are stable in the reduced $Fe^{2+}$ state unlike hemoglobin and myoglobin which are oxidized to the ferric ($Fe^{3+}$) state in air. Because DyPs and DHPs utilize the oxidized, $Fe^{3+}$ state to react with hydrogen peroxide and substrate, it is unlikely that NO-sensing heme proteins in the resting state could do the same type of peroxidase chemistry. In fact, oxidants like hydrogen peroxide inactivate mammalian sGC. Although bacterial NO-sensing heme proteins are more robust than mammalian sGC and could be developed as bioremediation catalysts, the differences in the heme coordination chemistry and redox properties make them less promising candidates.

## 1.6 HEME-BASED CO SENSORS (CooA and Homologs)

The first carbon monoxide-sensing heme protein, CooA, was discovered in the proteo-bacterium *Rhodospirillum rubrum*, and is now known to include a variety of CO-sensing heme proteins (Shelver et al. 1997, Roberts et al. 2005, Shimizu et al. 2015). In *R. rubrum*, a lake bacterium, carbon monoxide can be used as the sole energy source by initiating the expression of a CO oxidizing system (Kerby et al. 1995). The heme-based CO sensor CooA upregulates the CO-oxidizing system by binding to DNA in response to elevated levels of CO (Shelver et al. 1997). Although it was clear that CooA was a homodimeric heme protein when it was initially purified and characterized, it has taken years of struc-tural and spectroscopic studies to produce a more detailed mechanism of the heme coordination environment and CO sensing mechanism (Shelver et al. 1997).

Initial mutagenesis and spectroscopic studies of CooA from *R. rubrum* revealed that the heme of oxidized, $Fe^{3+}$ CooA is six-coordinate with a cysteine-thiolate as one of the two axial ligands bound to the heme iron (Reynolds et al. 1998). Upon the reduction of the heme iron to $Fe^{2+}$, the cysteine is replaced with a nearby histidine residue (Shelver et al. 1999). Carbon monoxide binding then displaces one of the two axial ligands causing CooA to be able to bind to DNA, upregulating the CO oxida-tion system. It was unclear which heme ligand was displaced by CO until a crys-tal structure revealed that the heme of reduced, $Fe^{2+}$ CooA has a novel N-terminal proline bound to the distal side of the heme iron along with the proximal histidine (Lanzilotta et al. 2000). Additional evidence showed that CO displaces the weak N-terminal proline ligand from the heme iron resulting in a structural change in the CooA protein that triggers DNA binding (Yamamoto et al. 2001).

Because the bacterial CooA proteins are stable and amenable to genetic manipu-lation, they are a potential target for making bioremediation catalysts. Although the oxidized, $Fe^{3+}$ state of CooA is stable, the fact that the heme iron is six-coordinate and fully saturated with a proximal histidine and an N-terminal proline, means that hydrogen peroxide and substrates cannot interact directly with the heme iron of ferric $Fe^{3+}$-CooA. The N-terminal proline can be removed using site-directed mutagenesis but the fact that there is a cysteine-thiolate bound to the heme iron in the oxidized, $Fe^{3+}$ state is problematic. Both the DyPs and the DHPs have a histidine as the proximal heme ligand. The change in proximal ligands has been shown to drastically change the type of reaction catalyzed by heme proteins. In fact, heme proteins with cysteine as the proximal heme ligand were originally found in the ubiquitous Cytochrome P450 fam-ily of heme enzymes that activate oxygen for substrate oxidation. Thus, it would be sur-prising for CO-sensing CooA to carry out the same type of peroxidase reactions as the DyPs or the DHPs, even with significant genetic alteration of the heme environment.

## 1.7 HEME-PAS PROTEINS (FixL and Others)

The heme-PAS family, named for the Per, Arnt, and Sym proteins that have a similar PAS structural fold, was one of the first gas sensing heme protein families discovered in nature (Gilles-Gonzalez and Gonzalez 2005, Shimizu et al. 2015). The FixL protein from *Sinorhizobium meliloti* was the first heme-PAS protein discovered and is a sym-biotic bacterium found in the root nodules of alfalfa plants (Gillis-Gonzalez et al. 1991,

Monson et al. 1992). There is a homolog FixL from *Bradyrhizobium japonicum* that is fully soluble, and therefore its structure has been more extensively characterized (Gilles-Gonzalez and Gonzalez 2005). FixL is also a part of the histidine kinase family, meaning that the protein has both a heme domain and a protein signaling kinase domain.

When the heme of FixL has bound $O_2$, the kinase domain is in the kinase "off" state with the kinase of FixL being inactive. Conversely, when the $O_2$ is not bound to the heme iron, in the deoxy state, the kinase domain is in the kinase "on" state with the kinase domain of FixL being active (Gilles-Gonzalez and Gonzalez 2005). In the deoxy state, FixL is an $\alpha_2$ homodimer that autophosphorylates its histidine kinase which then binds to, and phosphorylates, dimeric FixJ (Gilles-Gonzalez and Gonzalez 2005). Phosphorylated FixJ binds to DNA and activates the expression of the *nif* and *fix* genes that upregulate nitrogen fixation and micro aerobic respiration under low $O_2$ conditions in *S. meliloti*.

The heme of BjFixL is sandwiched between an N-terminal $\alpha$ helix and several $\beta$ sheets in a catcher's mitt or basket structure (Gilles-Gonzalez and Gonzalez 2005). There is a proximal histidine bound to the heme iron and an open site on the distal side of the heme to bind $O_2$, although the distal pocket does not contain a distal histidine like hemoglobin or myoglobin. Instead there are two conserved arginines, one that is on the distal side and other that interacts with a heme propionate. Researchers have proposed that the movement of these conserved arginine triggers a structural change in the heme domain of FixL which affects the C terminal kinase domain that lead to the kinase activity (Gilles-Gonzalez and Gonzalez 2005, Reynolds et al. 2009). There are also several conserved distal residues in the heme domain including a conserved isoleucine and a leucine.

How could heme-PAS proteins be used to degrade dyes and halophenols? Like the DyPs and the DHPs FixL proteins are stable in their oxidized, $Fe^{3+}$ states. In addition, FixL has an arginine on the distal side of the heme that is similar to the DyPs. To generate the aspartate-arginine distal pair found in the DyPs, either the conserved distal isoleucine or the leucine in FixL could be changed to an aspartate using site-directed mutagenesis to generate I251D or L236D. There is already a distal arginine present in the heme domain of FixL. This could result in a FixL protein with an aspartate-arginine distal pair that can degrade anthraquinone dyes like the DyPs. There are also several conserved tryptophan and tyrosine residues in the extended second coordination sphere that could facilitate electron transfer from the surface of FixL similar to DyPs. Further studies are needed to confirm whether FixL proteins can be turned into DyPs.

What about converting FixL into a dehalophenol peroxidase? The DHPs have a coordination environment more similar to globins like myoglobin and hemoglobin with a distal histidine and a distinctive distal coordination sphere. Although a distal histidine could be introduced into FixL using site-directed mutagenesis, the second coordination sphere of FixL is markedly different from DHPs. Thus, FixL appears to be a better model for the DyPs than the DHPs.

## 1.8  GLOBIN-COUPLED SENSORS (GCSs)

GCS proteins are a class of heme-based gas sensing proteins that, over the last two decades, have become recognized as encompassing proteins for all phyla of life (Gilles-Gonzalez and Gonzalez 2005, Shimizu et al. 2015). GCSs include the HemAT protein from *Halobacterium salinarum* and *Bacillus subtilis*, the *Escherichia coli* DOS protein, the histidine kinase AfGcHK from the soil bacterium *Anaeromyxobacter* and the GsGCS from *Geobacter sulfurreducens*. The GCS heme-based gas sensors sense oxygen and regulate a variety of important biological processes (Gilles-Gonzalez and Gonzalez 2005, Shimizu et al. 2015).

The bacterial GCSs have an N-terminal globin like fold containing α helices similar to hemoglobin and myoglobin that binds heme, and an enzymatic domain such as a kinase, phosphodiesterase, diguanylate cyclase, or a methyl accepting chemotaxis in the C terminal domain (Shimizu et al. 2015). Unlike the heme-PAS proteins, GCSs are typically activated by oxygen binding to the reduced, $Fe^{2+}$ heme iron. The structures of the oxidized, $Fe^{3+}$ states of HemeAT and GsGCS have been determined (Shimizu et al. 2015). The heme of HemeAT contains a proximal histidine ligand bound to the heme iron like hemoglobin and myoglobin., but with a distal tyrosine, instead of the distal histidine of hemoglobin. In contrast, the heme of GsGCS is in the oxidized, $Fe^{3+}$ state is six-coordinate with a distal histidine bound to the heme iron as well. These differences in heme coordination in the GCS family likely modulate their different sensitivities to oxygen binding.

Because of their globin heme-binding domain, the GCS heme proteins have potential for being modified to work as peroxidase catalysts in degrading toxic dyes and halophenols. The distal tyrosine could be changed to an aspartate to try and convert it into a dye-decolorizing peroxidase. However, the DYPs have an aspartate-arginine distal pair so a double mutant might be necessary. Slightly more promising is the possibility of converting HemeAT or other GCSs to DHPs by converting the distal tyrosine into a distal histidine like that found in the DHPs. While this could work to make an active dehalogenation catalyst, it is unclear whether the heme pocket of GCSs can do the same chemistry as DHPs. The fact that GCS proteins like HemeAT have similar globin-like folds and heme domains to DHPs makes them potential targets for making more efficient enzymes for bioremediation.

## 1.9 CONCLUSION

Water pollution is one of the biggest environmental concerns on our planet. This is the direct result of human-made pollutants including stable textile dyes and halophenols. However, the typical processes of using chemical or physical treatments can generate additional waste that is both expensive and harmful to the environment. Thus, researchers have searched for new green methods that are cheaper and more environmentally friendly to remove these toxic compounds. Fortunately, new heme peroxidase enzymes have been discovered that degrade anthraquinone dyes and halophenols that otherwise persist in the environment and lead to severe health complications.

However, DYPs have low catalytic activity which makes them less useful. In order to improve the catalytic properties of these heme peroxidases in bioremediation, scientists have sought to genetically engineer heme enzymes to more efficiently degrade these toxic contaminants. Because heme-based gas sensing proteins are ubiquitous in nature, have a diverse range of heme domains, and can be genetically manipulated, they are an ideal target for new bioremediation catalysts. This study examined the possibility of using four different families of heme-based gas sensor proteins as catalysts in dye pollution degradation: the NO-sensing H-NOX, the CO-sensing CooA, the heme-PAS family, and the GCS, globin-coupled sensors.

The heme-PAS family which contains the FixL protein and the GCS family which contains the Heme AT protein appear to be the best candidates for creating genetically engineered heme peroxidases that can degrade toxic anthraquinone dyes and halophenols. This is because they both contain heme domains that have some simlarity to DyPs and DHPs, where hydrogen peroxide and substrate bind, and catalysis occurs. Although the CooA and H-NOX families of heme-based gas sensing families have been well studied and can be genetically manipulated, their heme coordination environments and redox potentials make them less promising targets for peroxidase catalysts.

The use of heme enzymes from genetically engineered bacteria or fungi would be much cheaper and significantly more environmentally friendly than the traditional methods that have been used to degrade textile dyes (Routoula and Patwardhan 2020). These techniques are still being developed in the field but have shown promise in countries where water pollution from textile dyes is a significant problem, particularly, in Asia (Gita et al. 2017). The key will be to design new heme enzymes that are stable to pH changes, temperature changes, salinity, or other conditions that are present in industrially polluted water. Harnessing the power of genetically engineered heme enzymes to degrade textile dyes may turn heme proteins into the green catalysts of the future.

# REFERENCES

Chen, Y. P., Woodin, S. A., Lincoln, D. E., Lovell, C. R. 1996. An unusual dehalogenating peroxidase from the marine terebellid polychaete *Amphitrite ornata*. *Journal of Biological Chemistry*. 271 (9): 4609–4612.

D'Antonio, J., D'Antonio, E. L., Thompson, M. K., Bowden, E. F., Franzen, S., Smirnova, T., Ghiladi, R. A. 2010. Spectroscopic and mechanistic studies of dehaloperoxidase B from *Amphitrite ornata*. *Biochemistry*, 49: 6600–6616.

de Serrano, V., Chen, Z., Davis, M. F., Franzen, S. 2007. X-ray crystal structural analysis of the binding site in the ferric and oxyferrous forms of the recombinant heme dehaloperoxidase cloned from *Amphitrite ornata*. *Acta Crystallogr.* D63: 1094–1101.

Deinum, G., Stone, J. R., Babcock, G. T., Marletta, M. A. 1996. Binding of nitric oxide and carbon monoxide to soluble guanylate cyclase as observed with resonance Raman spectroscopy. *Biochemistry*. 35: 1540–1547.

Du, J., Huang, X., Sun, S., Wang, C., Lebioda, L., Dawson, J. H. 2011. *Amphitrite ornata* dehaloperoxidase (DHP): investigation of structural factors that influence the mechanism of halophenol dehalogenation using "peroxidase-like" myoglobin mutants and "myoglobin-like" DHP mutants. *Biochemistry*. 50: 8172–8180.

Field, M. J., Bains, R. K., Warren, J. J. 2017. Using an artificial tryptophan "wire" in cytochrome c peroxidase for oxidation of organic substrate. *Dalton Transactions.* 46: 11078–11083.

Franzen, S., Thompson, K. 2012. The dehaloperoxidase paradox. *Biochimica et Biophysics Acta*. 1824 (4): 578–588.

Ghaly, A., Ananthashankar, R., Alhattab, M., Ramakrishnan, V. V. 2013. Production, characterization and treatment of textile effluents: a critical review. *Journal of Chemistry Engineering Process Technology*. 5 (1): 182–199.

Gita, S, Hassan, A., Choudhury, T. G. 2017. Impact of textile dye waste on aquatic environments and its treatment. *Environment and Ecology.* 35 (3C): 2349–2353.

Godoy, F., Zenteno, P., Cerda, F., Gonzalez, B., Martinez, M. 1999. Tolerance to trichlorophenols in microorganisms from a polluted and a pristine site of a river. *Chemosphere* 38 (3): 655–662.

Gilles-Gonzalez, M.-A., Ditta, D. S., Helinski, D. R.1991. A haemoprotein with kinas activity encoded by the oxygen sensor of Rhizobium *meliloti. Nature* 350: 170–172.

Gilles-Gonzalez, M.-A., Gonzalez, G. 2005. Heme-based sensors: defining characteristics, recent developments, and regulatory hypotheses. *Journal of Inorganic Biochemistry.* 99: 1–22.

Guo, Y., Lou, X., Fang, C., Xiao, D., Wang, Z., Lui, J. 2013. Novel photo-sulfite system: toward simultaneous transformations of inorganic and organic pollutants. *Environmental Science and Technology.* 47: 11174–11181.

Kerby, R. L., Ludden, P. W., Roberts, G. P. 1995. Carbon-monoxide dependent growth of *Rhodospirillum rubrum Journal of Bacteriology.* 177: 2241–2244.

Kim, S. J., Shoda, M. 1999. Purification and characterization of a novel peroxidase from *Geotrichum candidum* Dec 1 involved in decolorization of dyes. *Applied and Environmental. Microbiology.* 65: 1029–1035.

Lanzilotta, W. N., Schuller, D. J., Thorsteinsson, M. V., Kerby, R. L. Roberts, G. P., Poulos, T. L. 2000. Structure of the CO sensing transcription activator CooA. *Natural Structural Biology.* 7: 876–880.

Lellis, B., Fávaro-Polono, C. J., Pamphille, J. C., Polonio, J. C. 2019. Effects of textiles on health and the environment and bioremediation potential of living organisms. *Biotechnology Research and Innovation.* 3: 275–290.

Lin, Y.-W. 2019. Rational design of heme enzymes for a greener future. *Biotechnology and Applied Biochemistry.* 67: 484–494.

Linde, D., Ruiz-Dueñas, F. J., Fernández-Fueyo, E. 2015. Basidiomycete DyPs: Genomic diversity, structural–functional aspects, reaction mechanism and environmental significance. *Archives of Biochemistry and Biophysics.* 574: 66–74.

Monson, E. K., Weinstein, M., Ditta, G. S., Helinski, D. R. 1992. The FixL protein of rhizobium meliloti can be separated into a heme-binding oxygen-sensing domain and a functional C-terminal kinase domain. *Proceedings of the National Academy of Science USA* 89: 4280–4284.

Osborne, R. L., Coggins, M. K., Walla, M., Dawson, J. H. 2007. Horse heart myoglobin catalyzes the $H_2O_2$-dependent oxidative dehalogenation of chlorophenols to DNA-binding radicals and quinones. *Biochemistry* 46:9823–9829.

Pfanzagl, V., Nys, K., Bellei, M., Michlits, H, Mlynek, G. Battistuzzi, G, Djinovi-Carugo, K, Doorslaer, S.-V., Furtmüller, P. G., Hofbauer, F., Obinger, C. 2018. Roles of distal aspartate and arginine of B-class dyedecolorizing peroxidase in heterolytic hydrogen peroxide cleavage. *Journal of Biological Chemistry.* 293 (38): 14823–14838.

Poulos, T. J. 2014. Heme structure and function. *Chemical Reviews.* 114: 3919–3962.

Reynolds, M. F., Burstyn, J. N. 2001. Mechanism of Activation of Soluble Guanylate Cyclase by NO: Allosteric Regulation Through Changes in Heme Coordination Geometry. *Nitric Oxide: Biology and Pathobiology, Ed., L. Ignarro,* 381–389.

Reynolds, M. F., Shelver, D., Kerby, R. L., Parks, R. B. 1998. EPR and electronic absorption spectroscopies of the CO-sensing CooA protein reveal a cysteine-ligated low-spin ferric heme. *Journal of American Chemical Society.* 120: 9080–9081.

Reynolds, M. F., Ackley, L., Blizman, A., Lutz, Z., Manoff, D. R., Miles, M., Pace, M, Patterson, J., Pozzessere, N., Saia, K., Sato, R., Smith, D., Tarves, P., Weaver, M., Sieg, K., Lukat-Rodgers, G. S., Rodgers, K. R. 2009. Role of Conserved Fα-helix Residues in the Native Fold and Stability of the Kinase-Inhibited Oxy State of the Oxygen-sensing FixL Protein from Sinorhizobium meliloti. *Archives of Biochemistry and Biophysics.* 485: 150–159.

Roberts, G. P., Kerby, R. L., Youn, H., Conrad, M. 2005. CooA, a paradigm for gas sensing regulatory proteins. *Journal of Inorganic Biochemistry*, 99: 280–292.

Routoula, E., Patwardhan, S. V. 2020. Degradation of anthraquinone dyes from effluents: a review focusing on enzymatic dye degradation with industrial potential. *Environmental Science and Technology*. 54: 647–664.

Shimizu, T., Huang, D., Yan, F., Stranava, M., Bartosova, M., Fojtikova, V., Matrinkova, M. 2015. Gaseous $O_2$, NO and CO in signal transduction: structure and function relationships of heme-based gas sensor and heme-redox sensor. *Chemical Reviews*. 115: 6491–6533.

Shelver, D., Kerby, He, Y., Roberts. 1997. CooA, a CO-sensing transcription factor form *Rhodospillum rubrum*, is a CO-binding heme protein. *Proceedings of the National Academy of Science USA*. 94: 11216–11220.

Shelver, D., Thorsteinsson, M. V., Kerby, R. L., Chung, S.-Y., Roberts, G. P. Reynolds, M. F. Parks, R. B., Burstyn, J. N. 1999. Identification of two important heme site residues (cysteine 75 and histidine 77) in CooA, the CO-sensing transcription factor of Rhodospirillum rubrum. *Biochemistry*. 38: 2669–2678.

Singh, R., Eltis, L. D. 2015. The multihued palette of dye-decolorizing peroxidases. *Archives of Biochemistry and Biophysics*. 574: 55–66

Shrestha, R., Chen, X., Ramyar, K. X., Hayati, Z., Carlson, E. A., Bossmann, S. H., Song, L., Geisbrecht, B. V., Li, P. 2016. Identification of surface-exposed protein radicals and a substrate oxidation site in A-class dye-decolorizing peroxidase from *Thermomonospora curvata*. *ACS Catalysis*. 6: 8036–8047.

Waldman, S. A., Murad, F. 1987. Cyclic GMP synthesis and function. *Pharmacological Review*. 39: 163–196.

Wang, C., Lovelace, L. L., Sun, S., Dawson, J. H., Lebioda, L. 2013. Complexes of dual-function hemoglobin/dehaloperoxidase with substrate 2,4,6-trichlorophenol are inhibitory and indicate binding of halophenol to compound I. *Biochemistry* 52, 6203–6210.

Watkins, D. W. (2017) Construction and *in vivo* assembly of a catalytically proficient and hyperthermostable de novo enzyme. *Nature Communications* 358: 1–9.

Wolin, M. S., Wood, K. S., Ignarro, L. J. 1982. Guanylate cyclase from bovine lung. *Journal of Biological Chemistry*. 257: 13312–13320.

Yamamoto, K., Ishikawa, H., Takahashi, S., Ishimori, K., Morishima, I., Nakajima, H., Aono, S. 2001. Binding of CO at the $Pro^2$ side chain is crucial for the activation of CO-sensing transcriptional activator CooA. *Journal of Biological Chemistry*. 276: 11473–11476.

Yin, L.-L., Yuan, H., Liu, C., He, B., Gao, S.-Q., Wen, G.-B., Tan, X., Lin, Y.-W. 2018. A rationally designed myoglobin exhibits a catalytic dehalogenation efficiency more than 1000-fold that of a native dehaloperoxidase. *ACS Catalysis*. 8: 9619–9624.

Yoshida, T., Sugano, Y. 2015. A structural and functional perspective of DyP-type peroxidase family. *Archives of Biochemistry and Biophysics*. 574: 49–55.

Yu, A. E., Hu, S. Z., Spiro, T. G., Burstyn, J. N. 1994. Resonance Raman spectroscopy of soluble guanylyl cyclase reveals displacement of distal and proximal heme ligands by NO. *Journal of the American Chemical Society*. 116: 4117–4118.

Zhang, P., Xu, J., Want, X.-J., He, Bo, Gao, S.-Q., Lin, Y.-W. 2019. The third generation of artificial dye-decolorizing peroxidase rationally designed in myoglobin. *ACS Catalysis*. 9: 7888–7893.

# 2 Organic Farming with Residual Microbial Consortia and Its Potential in Sustainable Agriculture Production

*Suhana Sultana, Tamanna Sultana,*
*Arup Kumar Mitra, and Xavier Savarimuthu, SJ*

## CONTENTS

## 2.1 INTRODUCTION: PLANT-MICROBE INTERACTION IN NATURE AND ITS SIGNIFICANCE IN AGRICULTURE

The growth and development of plants is highly dependent on plant and microbe complexes which are known as holobiomes (O'Malley 2017). Symbiotic microbes, along with endophytic microorganisms, play a vital role in enhancing nutrient uptake, subsequent assimilation of nutrients, rate of photosynthesis, and help to enhance disease or pest resistance (Hacquard 2015; Johnson et al. 2018; Jones et al. 2007). Such bio-stimulants can be considered a potential tool to reduce the dependence on chemical fertilizers and other chemical treatments, which often have a negative impact on the environment. Plant products, certain proteins, and bioactive extracts are a part of bio-stimulants and can improve plant growth and help in sustainable crop production. The associated mechanisms have been shown in Figure 2.1.

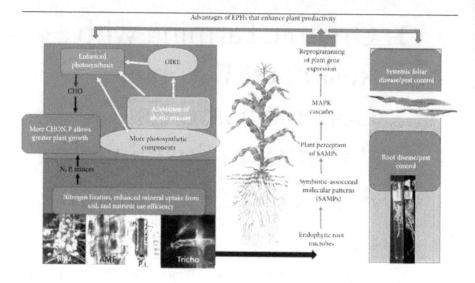

**FIGURE 2.1**  A summary of mechanisms and systems associated with endophytic microorganisms by which they increase plant productivity.

Certain single strain microbial bio-stimulants have been found to increase the rate of nitrogen fixation in leguminous plants (Boogerd and Van Rossum 1997; Markmann and Parniske 2009; Wu et al. 2018). However, there is an increasing tendency to use a combination of microbial and non-microbial bio-stimulants to gain benefits from synergistic interactions. Rhizosphere signals and eco-physiological responses of the host plant selectively activate different members of the inoculated microbial communities to exert an advantageous impact on the growth of crops (Samuels and Hebbar 2015; Schmidt et al. 1994; Tahat et al. 2014). Genetically diverse groups of microbial strains can differentially and selectively adapt to factors like temperature, soil pH, and moisture content of the soil. For example, in the absence of soluble forms of phosphorus in the soil, phosphate solubilizing bacteria, and phosphate mineralizers are activated in response to the growth of pathogenic fungi.

Crops treated with microbial and non-microbial bio-stimulants have certain advantages over crops grown by conventional methods:

1. Improved root system leading to improved growth of the plant body (Yanni et al. 1997, 2001, 2011, 2016)
2. Enhanced crop productivity and yield (South et al. 2019; Chi et al. 2005)
3. Better and efficient uptake and assimilation of nutrients, as in Rhizobiaceae's nitrogen fixation in leguminous plants (Boogerd et al. 1997)
4. Improved resistance toward insects or pests and toward abiotic stresses like salinity, drought, and temperature fluctuations (Das et al. 2015; Porcel et al. 2011, 2015)
5. Increased rate of gaseous exchange and photosynthesis (Khan et al. 2008)

## 2.2   WHAT ARE PLANT BIO-STIMULANTS?

Bio-stimulants applied to plants have been found to increase crop yield and improve the nutritional content and quality of the products. Sustainable agricultural practices include the use of bio-stimulants to increase crop productivity and restore the equilibrium in the soil ecosystem. Bio-stimulants improve the uptake of nutrients and tolerance to biotic and abiotic stress (Gill et al. 2016), enhancing the quality of crops. Commercial bio-stimulants are typically composed of a mixture of certain amino acids, proteins, humic acid, fulvic acid, plant extracts, polymers, and elements like aluminum, sodium, cobalt, silicon, and bacteria or fungi (Yakhin et al. 2017; Akhtar et al. 2015). The term "bio" may seem ambiguous in certain cases because some of the listed components are not biological in character; however, non-organic factors can act as catalysts for the biological reactions and processes, and constitutively regulate plant growth and development.

## 2.3   REGULATORY LEGISLATION PROTOCOL – PLANT PROTECTION PRODUCTS VERSUS PLANT FERTILIZERS: AN OVERVIEW

The protocol for registration of agricultural products in European countries is broadly divided into two main categories:

1. Plant Protection Products
2. Plant Fertilizers

Plant protection products (PPPs), composed of beneficial microbes and certain chemicals, as defined in Council Regulation (EC)1, provide protection to plants against opportunistic pathogens and harmful microbes and other organisms, regulate plant metabolism and reactions, and destroy redundant parts of the plant. The process of PPP registration is time consuming and may not be applicable for all bio-stimulants (Jardin 2015). Hence, PPP producers often look for an alternative procedure that would be suitable for registration of plant bio-stimulants, while simultaneously requiring less time and required costs to get the products registered and recognized in the market.

To date, regulatory processes for plant bio-stimulants are yet to be established, and this is a topic of discussion in the United States and EU. It is essential to consider incorporating registration policies for microbial and non-microbial bio-stimulants, microbial strains and consortia, and their metabolic products. Several microorganisms, such as *Trichoderma* spp., have been registered as PPPs and categorized as microbial biological control agents (MBCA). Also, there are plant bio-stimulants, such as arbuscular mycorrhizal fungi (AMF), that are capable of inducing resistance in the plant body against pest attack (Parniske 2008; Strack et al. 2003; Weindling et al. 1932) and other diseases (Cameron et al. 2013; Lorito and Woo 2015). There is a need for the creation of a dedicated registration procedure for microbe-based bio-stimulants to form a part of sustainable agricultural practice. This registration procedure will help in the development of an all-inclusive method to effectively regulate the use of agricultural products, along with biofertilizers, biopesticides, and bio-stimulants.

## 2.4  PLANT GROWTH-PROMOTING MICROBES IN DEVELOPING AND FORMULATING BENEFICIAL MICROBIAL CONSORTIA FOR SUSTAINABLE AGRICULTURE

Plant growth-promoting rhizobacteria (PGPR) are associated with the promotion of plant growth and include species like *Pseudomonas, Azotobacter, Bacillus,* and *Azospirillum*. They improve the availability of nutrients and minerals in the soil, nutrient uptake and utilization by the plant, and support nitrogen fixation. PGPM also include fungal species, like *Gigaspora, Rhizophagus (Glomus),* and *Laccaria*, that form a symbiotic relationship with the majority of vascular plants (Alaux et al. 2018). These fungal species enhance the water absorption capacity of the plant, regulate the gaseous exchange process of the plant with the environment, and can develop the plant's ability to tolerate abiotic stresses. The fungus *Trichoderma* is also an example of PGPM (Furch et al. 2016). Owing to its multiple beneficial effects on plants, it is an active ingredient in hundreds of agricultural products commercialized worldwide (Harman et al. 2004; Woo et al. 2014).

Two examples of PGPM that have been widely studied are *Trichoderma* and *Azotobacter*. *Trichoderma* was previously considered to possess only biopesticidal activity (Klein and Eveleigh 1998), but *Trichoderma* was later found to have multiple beneficial effects on plants. Various studies have confirmed *Trichoderma's* role as bio-stimulants, biofertilizers, and a factor in improved crop resistance against abiotic stresses and pest attack (Bae et al. 2011; Harman 2000; Neumann et al. 2006). Symbiotic interaction between *Trichoderma* and the plant can bring out *Trichoderma's* plant growth-promoting effect (Harman et al. 2004; Fontenelle et al. 2011; Harman et al. 2004; Vinale et al. 2008; Shoresh et al. 2010; Studholme et al. 2013; Lorito and Woo 2015). Under specific conditions, *Trichoderma* may induce a stimulus in the plant (i.e., priming), leading to a quick response to invasion by pathogens, which eventually can form a part of the development of a systemic acquired resistance (SAR) and/or induced systemic resistance (ISR) (Rubio et al. 2014; Martínez et al. 2017; Hossain et al. 2017; Manganiello et al. 2018; Berch et al. 2005). Furthermore, experimental outcomes and on-field application of *Trichoderma* on several kinds of crops suggest the alleviation of symptoms in response to abiotic diseases' (e.g., pH, moisture, temperature, nutrients) post-treatment methodologies (Mastouri et al. 2012; Brotman et al. 2013; Fiorentino et al. 2018). The indicators of improved plant health include enhanced seed germination, well-developed root and shoot systems, increased chlorophyll content (Ort et al. 2015), increased flowering, and fruiting in the plant (Harman et al. 2004; Hermosa et al. 2012; Studholme et al. 2013; Mendoza et al. 2018). Specifically, targeted modification in the root system increases the absorption area, resulting in an improved uptake of nutrients, which leads to the development of an effective translocation system (Panstruga and Kuhn 2015). Consequently, NPK uptake efficiency increases the availability of more micronutrients and contributes directly to increased biomass content (Samolski et al. 2012).

The phosphate solubilization ability and potential of *Trichoderma* contributes to the PGP effect (Altomare et al. 1999, Shoresh and Harman 2008), which is primarily attained by siderophore production and production of secondary metabolites (Vinale et al. 2009, 2013, 2014) or through modifying the concentrations of plant hormones, such as ethylene and auxin (Hermosa et al. 2013; Contreras-Cornejo et al. 2015), that aid in plant growth and development. Research and observations indicate that *Trichoderma* spp. alone have the ability to produce more than 250 metabolic products, including cell wall degrading enzymes, peptides and other essential protein-aceous products, and some important secondary metabolites (Sivasithamparam and Ghisalberti 1998; Harman et al. 2004; Morán-Diez et al. 2009; Lorito et al. 2010; Keswani et al. 2014; Ruocco et al. 2015). It has huge economic importance especially in rural India as shown in Figure 2.2. Potential bioactivity has been noted in many of these compounds that have the ability to regulate the plant's response to other microbes, which is achieved through restructuring the plant's defense mechanisms. Parallel plant growth and development ensure the development of a proper root system for the enhanced availability of micronutrients that are essential for plant growth and developing defense mechanisms within the plant (Sivasithamparam and Ghisalberti 1998; Vinale et al. 2009, 2014; Lombardi et al. 2018). Synergistic biocontrol effects have been observed in many consortia of diverse strains, secondary metabolites, and bioactive compounds derived from *Trichoderma*

**FIGURE 2.2** Packing of *Trichoderma viride* biofertilizers in a village production center in Tamil Nadu (India).

(Zachow et al. 2016) and other microorganisms or plants, suggesting an array of new pathways for developing sophisticated products in the form of bio-stimulants that can be used in sustainable agriculture practices.

The nitrogen fixing free-living bacteria *Azotobacter* is known to be tolerant toward various stress factors. *Azotobacter's* nitrogen fixing ability directly influences agroecosystems, leading to an increased soil level and availability of this vital element for uptake by plants. *Azotobacter* is tolerant to abiotic stresses, like drought and salinity, and is capable of forming heat-resistant cysts, constituting an inoculant with a good shelf-life (Inamdar et al. 2000; Vacheron et al. 2013; Berg et al. 2014; Viscardi et al. 2016).While conferring stress resistance to the plant, *Azotobacter* can simultaneously positively interact with other microorganisms in agroecosystems (Babalola 2010; Ahmad et al. 2011; Berendsen et al. 2012; Bhattacharyya and Jha 2012; Gaiero et al. 2013; Philippot et al. 2013). *Azotobacter* is an active ingredient in numerous commercial biofertilizer products, often in association with fungi, *Actinomycetes*, and some other bacteria, such as bacilli (EBIC 2013). *Azotobacter* strains also secrete plant growth-promoting substances, including vitamins, phytohormones, and antifungal metabolites, which help in the growth and differentiation of plant tissues. The process of solubilization of phosphate in soil (Hariprasad and Niranjana 2009; Rojas-Tapias et al. 2012; Wani et al. 2013) and mobilization of iron (Rizvi and Khan 2018) have been successfully inferred through a series of lab-based and on-field experimental observations, and promising results were obtained when abiotic stress conditions were provided (Viscardi et al. 2016; Van Oosten et al. 2018). Another interesting aspect is that plant productivity and yield is notably enhanced by *Azotobacter*-mediated synthesis of enzymes, such as catalase (CAT), superoxide dismutase (SOD), proline, and elevated levels of 1-aminocyclopropane-1-carboxylate (ACC) activity (Glick 2014). These results have been further confirmed in a wide variety of crops, like tomato, rice, wheat, maize, and sorghum (Rojas-Tapias et al. 2012; Inamdar et al. 2000; Long et al. 2018; Di Stasio et al. 2017; Viscardi et al. 2016). Recently, a study conducted by Barra et al. (2016) further validated the significance of the enzyme ACC deaminase (ACCd) activity and the production of indole-3acetic acid (IAA) responsible for withstanding salt stress in crops inoculated with *Azotobacter* strains that are stress tolerant. Simultaneously, Hermosa et al. (2012) indicated a model system that established the vital role of IAAs and ACCd produced by *Trichoderma* in controlling and maintaining an equilibrium state between plant defense system and plant growth development process.

## 2.5   CONCLUSION: AGRICULTURAL PROBIOTICS — A PROMISING REMEDY COMPRISING OF FORMULATION OF NOVEL MICROBIAL CONSORTIA FROM RESIDUAL SOIL MICROBIOTA TO ENHANCE PGP EFFICACY

The emerging concept of rhizosphere engineering proposes the inoculation or application of specific microbial strains to improve and restructure the networks prevailing in the soil. This process facilitates efficient and beneficial microbes increasing soil fertility by exerting functional recovery in the soil system, achieved through their metabolic processes and microbial products synthesizing within the soil and plant body (Stringlis et al. 2018; Wallenstein 2017; Ruzzi and Aroca 2015; Shi et al. 2016), and by restoring the original microbiome that was lost or reduced by detrimental agricultural methodologies (Leff et al. 2016; Perez-Jaramillo et al. 2016). Adopting the practices of applying residual microbial consortia will efficiently encompass the regulation of various metabolic processes:

1. Phosphate solubilization
2. Conversion of insoluble forms of minerals and elements to their soluble forms, along with making them available to the plants
3. Elevating nitrogen fixation ability
4. Producing considerable amounts of essential phytohormones, enzymes, and siderophores. These will provide optimum growth and development to the plant body while simultaneously conferring protection to the plant against opportunistic pathogens and abiotic stresses, such as high temperature, drought, extreme salt concentrations (Ashraf et al. 2004; Compant et al. 2005; Gopalakrishnan et al. 2015; Van Oosten et al. 2017), and toxicity due to accumulation of heavy metals, hazardous chemicals, and pesticides (Ventorino et al. 2014)

We have designed and formulated a microbial consortia based on our lab research as shown in Figures 2.3 and 2.4. While an in-depth knowledge is required in this field of study, there is satisfactory experimental evidence that has successfully demonstrated the ability of microbial consortia, comprising of rhizospheric soil bacteria and fungi, in establishing a competent association with the naturally prevailing soil microbiota (Hardoim et al. 2015; de Vries and Wallenstein 2017; Bonanomi et al. 2017). Novel microbial communities are often a result of interaction between different bacteria and fungi (Ahmad et al. 2011; Berg et al. 2014; Lugtenberg 2015;

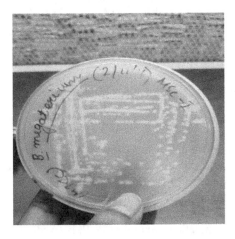

**FIGURE 2.3** *Bacillus megaterium* isolated from a rhizospheric soil sample (left) and a bacterial consortium (right) developed.

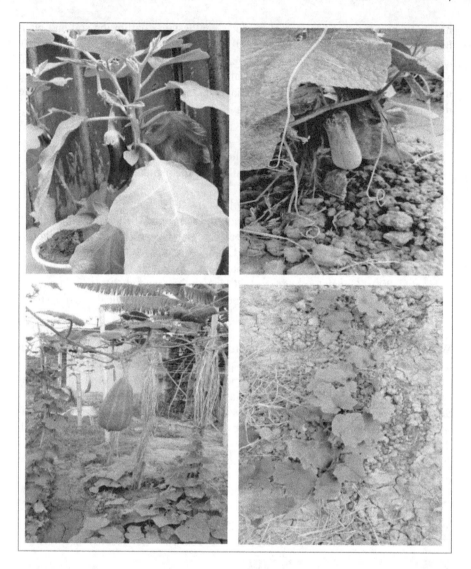

**FIGURE 2.4**   Increased productivity in various crops observed using microbial consortia as biofertilizers in a field trial at Purulia.

Mallock 2018). Applying a combination of different strains of microbes can bring out new plant growth-promoting effects better than those obtained by using single strain (Wargo and Hogan 2006; Hacquard et al. 2015). Hence, understanding the synergistic interactions between the constituent microbes, their beneficial effects on the plant body, and their effects on the adjacent environment and on the ecosystem as a whole is essential in developing proper formulation mixture of PGPMs (Andrews and Andrews 2017). Additionally, checking the technical viability and cost effective-ness of the products is important from the users' perspectives (Berendsen et al. 2012;

Berg et al. 2014; Lugtenberg 2015; Yakhin et al. 2017; Kong et al. 2018). Various detailed studies on *Trichoderma* and *Azotobacter* show that they can function in a complementary fashion when present in a plant growth-promoting consortium. Moreover, the *Trichoderma-Azotobacter* consortia may be supplemented with organic compounds, metabolic products of certain animals, proteins, and polymers to obtain an effective formulation that will promote plant growth and development. This requires detailed studies based on the genome to understand the mechanisms of the synergistic interactions of the members of the consortium to develop improved bio-stimulants (Bell et al. 2015; Fiorentino et al. 2018; Ventorino et al. 2018). As more and more countries are considering the idea of converting from conventional methods of farming to sustainable organic farming methods, the demand for bio-fertilizers, biopesticides, and other organic tools is increasing on a global scale (Woo et al. 2014; Lugtenberg 2015).

In recent years, a consistent growth of about 10% per year has been observed in the biofertilizer market (EBIC 2013). A deeper understanding of the microbial interactions in the rhizosphere and plant-microbe interactions will help in the development of novel and effective bio-stimulant formulations that will ensure improved crop yield. Microorganisms like *Trichoderma* and *Azotobacter*, and some related strains, can be used in a combination to enhance plant growth and crop yield. These resulting formulations can be considered a sustainable approach toward developing a viable method of organic farming.

## REFERENCES

Ahmad, I., M. S. A. Khan, F. Aqil, and M. Singh. 2011. "Microbial applications in agriculture and the environment: a broad perspective." In *Microbes and Microbial Technology: Agricultural and Environmental Applications.* Edited by I. Ahmad, F. Ahmad, and J. Pichtel, 1–27. New York, NY: Springer.
Akhtar, M. S., J. Panwar, S. N. A. Abdullah, Y. Siddiqui, M. K. Swamy, and S. Ashkani. 2015. "Biocontrol of plant parasitic nematodes by fungi: efficacy and control strategies." In *Organic Amendments and Soil* Soil Suppressiveness in Plant Disease Management, edited by M. K. Meghvansi and A. Varma, 46: 219–247.
Alaux, P. -L., V. César, F. Naveau, S. Cranenbrouck, and S. Declerck. 2018. "Impact of *Rhizophagus irregularis* MUCL41833 on disease symptoms caused by *Phytophthora infestans* in potato grown under field conditions." *Crop Protection* 107: 26–33.
Altomare, C., Norvell W. A., Bjorkman T., and G. E. Harman. 1999. "Solubilization of phosphates and micronutrients by the plant-growth-promoting and biocontrol fungus *Trichoderma harzianum* Rifai 1295-22." *Applied Environmental Microbiology* 65: 2926–2933.
Andrews, M., and M. E. Andrews. 2017. "Specificity in legume Rhizobia symbioses." *International Journal of Molecular Sciences* 18 (4): 705.
Ashraf, M., S. Hasnain, O. Berge, and T. Mahmood. 2004. "Inoculating wheat seedlings with exopolysaccharide-producing bacteria restricts sodium uptake and stimulates plant growth under salt stress." *Biology and Fertility of Soils* 40: 157–162. doi: 10.1007/s00374-004-0766-y.
Bae, H., D. P. Roberts, H-S. Lim, M. D. Strem, S.-C. Park, C.-M. Ryu, R. L. Melnick, and B. A. Bailey. 2011. "Endophytic *Trichoderma* Isolates from tropical environments delay disease onset and induce resistance against *Phytophthora capsici* in hot pepper using multiple mechanisms." *Molecular Plant-Microbe Interactions* 24 (3): 336–351.

Barra, P. J., Inostroza N. G., Acuña J. J., Mora M. L., Crowley D. E., and M. A. Jorquera. 2016. "Formulation of bacterial consortia from avocado (*Persea americana* Mill.) and their effect on growth, biomass and superoxide dismutase activity of wheat seedlings under salt stress." *Applied Soil Ecology* 102: 80–91. 10.1016/j.apsoil.2016.02.014

Berch, S. M., H. B. Massicotte, and L. E. Tackaberry. 2005. "Re-publication of a translation of 'The vegetative organs of Monotropa hypopitys L.' published by F. Kamienski in 1882, with an update on *Monotropa mycorrhizas*." *Mycorrhiza* 15 (5): 323–332.

Berendsen, R. L., Pieterse C. M., and P. A. Bakker. 2012. "The rhizosphere microbiome and plant health." *Trends Plant Science* 17: 478–486. 10.1016/j.tplants.2012.04.001

Berg, G., Grube M., Schloter M., and K. Smalla. 2014. "Unraveling the plant microbiome: looking back and future perspectives." *Frontier Microbiology* 5: 148. 10.3389/fmicb.2014.00148

Bhattacharyya, P. N., and D. K. Jha. 2012. "Plant growth-promoting rhizobacteria (PGPR): emergence in agriculture." *World Journal of Microbiology Biotechnology* 28: 1327–1350. 10.1007/s11274-011-0979-9

Bonanomi, G., Ippolito F., Cesarano G., Vinale F., Lombardi N., and A. Crasto, et al. 2017. "Biochar chemistry defined by13C-CPMAS NMR explains opposite effects on soilborne microbes and crop plants." *Applied Soil Ecology* 124: 351–361. 10.1016/j.apsoil.2017.11.027

Boogerd, F., and D. Van Rossum. 1997. "Nodulation of groundnut by *Bradyrhizobium*: a simple infection process by crack entry." *FEMS Microbiology Reviews* 21 (1): 5–27.

Brotman, Y., Landau U., Cuadros-Inostroza Á., Tohge T., Fernie A. R., and I. Chet, et al. 2013. "*Trichoderma*-plant root colonization: escaping early plant defense responses and activation of the antioxidant machinery for saline stress tolerance." *PLoS Pathogens* 9: e1003221. 10.1371/journal.ppat.1003221

Cameron, D. D., Neal A. L., van Wees S. C., and J. Ton. 2013. "Mycorrhiza-induced resistance: more than the sum of its parts?" *Trends Plant Science* 18: 539–545. 10.1016/j.tplants.06.004

Chi, F., S. -H. Shen, H. -P. Cheng, Y. -X. Jing, Y. G. Yanni, and F. B. Dazzo. 2005. "Ascending migration of endophytic rhizobia, from roots to leaves, inside rice plants and assessment of benefits to rice growth physiology." *Applied and Environmental Microbiology* 71 (11): 7271–7278.

Compant, S., Duffy B., Nowak J., Clément C., and E. A. Barka. 2005. "Use of plant growth-promoting bacteria for biocontrol of plant diseases: principles, mechanisms of action, and future prospects." *Applied Environmental Microbiology* 71: 4951–4959. 10.1128/AEM.71.9.4951-4959.2005

Contreras-Cornejo, H. A., López-Bucio J. S., Méndez-Bravo A., Macías-Rodríguez L., Ramos-Vega M., and Á. A. Guevara-García, et al. 2015. "Mitogen-activated protein kinase 6 and ethylene and auxin signaling pathways are involved in *Arabidopsis* root-system architecture alterations by *Trichoderma atroviride*." *Mol. Plant-Microbe Interact.* 28: 701–710. 10.1094/MPMI-01-15-0005-R

Das, S. N., J. Madhuprakash, P. V. S. R. N. Sarma, P. Purushotham, K. Suma, K. Manjeet, S. Rambabu, N. E. El Gueddari, B. M. Moerschbacher, and A. R. Podile. 2015. "Biotechnological approaches for field applications of chitooligosaccharides (COS) to induce innate immunity in plants." *Critical Reviews in Biotechnology* 35 (1): 29–43.

de Vries, F. T., and M. D. Wallenstein. 2017. "Below-ground connections underlying aboveground food production: a framework for optimising ecological connections in the rhizosphere." *Journal of Ecology* 105: 913–920. 10.1111/1365-2745.12783

Di Stasio, E., Maggio A., Ventorino V., Pepe O., Raimondi G., and S. De Pascale. 2017. "Free-living ($N_2$)-fixing bacteria as potential enhancers of tomato growth under salt stress." *Acta Horticulturae* 1164: 151–156. 10.17660/ActaHortic.2017.1164.1

Fontenelle, A. D. B., Guzzo S. D., Lucon C. M. M., and R. Harakava. 2011. "Growth promotion and induction of resistance in tomato plant against *Xanthomonas euvesicatoria* and *Alternaria solani* by *Trichoderma* spp." *Crop Protection* 30: 1492–1500. 10.1016/j.cropro.2011.07.019

Furch, S., S. Gruber, R. Bansal, and P. K. Mukherjee. 2016. "Secondary metabolism in Trichoderma—chemistry meets genomics." *Fungal Biology Reviews* 30 (2): 74–90.

Gaiero, J. R., McCall C. A., Thompson K. A., Dayu N. J., Best A. S., and K. E. Dunfield. 2013. "Inside the root microbiome: bacterial root endophytes and plant growth promotion." *American Journal of Botany* 100: 1738–1750. 10.3732/ajb.1200572

Gill, S. S., R. Gill, D. K. Trivedi, N. A. Anjum, K. K. Sharma, M. W. Ansari, A. A. Ansari, A. K. Johri, R. Prasad, E. Pereira, A. Varma, and N. Tuteja. 2016. "*Piriformospora indica*: potential and significance in plant stress tolerance." *Frontiers in Microbiology* 7: 332.

Glick, B. R. 2014. "Bacteria with ACC deaminase can promote plant growth and help to feed the world." *Microbiology Research* 169: 30–39. 10.1016/j.micres.2013.09.009

Gopalakrishnan, S., Sathya A., Vijayabharathi R., Varshney R. K., Gowda C. L. L., and L. Krishnamurthy. 2015. "Plant growth promoting rhizobia: challenges and opportunities." *3 Biotech* 5: 355–377. 10.1007/s13205-014-0241-x

Hacquard, S. 2015. "Disentangling the factors shaping microbiota composition across the plant holobiont." *New Phytologist* 209 (2): 454–457.

Hardoim, P. R., van Overbeek L. S., Berg G., Pirttilä A. M., Compant S., and A. Campisano, et al. 2015. "The hidden world within plants: ecological and evolutionary considerations for defining functioning of microbial endophytes." *Microbiology Molecular Biological Review* 79: 293–320. 10.1128/MMBR.00050-14

Hariprasad, P., and S. R. Niranjana. 2009. "Isolation and characterization of phosphate solubilizing rhizobacteria to improve plant health of tomato." *Plant Soil* 316: 13–24. 10.1007/s11104-008-9754-6

Harman, G. E. 2000. "Myths and dogmas of biocontrol changes in perceptions derived from research on *Trichoderma harzinum* T-22." *Plant Disease* 84 (4): 377–393.

Harman, G. E., C. R. Howell, A. Viterbo, I. Chet, and M. Lorito. 2004. "*Trichoderma* species—opportunistic, avirulent plantsymbionts." *Nature Reviews Microbiology* 2 (1): 43–56.

Harrison, M. J. 1997. "The arbuscular mycorrhizal symbiosis: an underground association." *Trends in Plant Science* 2 (2): 54–60.

Hossain, M. M., Sultana F., and S. Islam. 2017. "Plant Growth-Promoting Fungi (PGPF): Phytostimulation and induced systemic resistance." In *Plant-Microbe Interactions in Agro-Ecological Perspectives, Volume 2: Microbial Interactions and Agro-Ecological Impacts.* Edited by D. Singh, H. Singh, and R. Prabha, 135–191. Singapore: Springer. 10.1007/978-981-10-6593-4

Inamdar, S., Kanitkar R. U., and M. G. Watve. 2000. "Longevity of *Azotobacter* cysts and a model for optimization of cyst density in liquid bioinoculants." *Current Science* 78: 719–721.

Johnson, J. M., J. Thürich, and E. K. Petutschnigetal. 2018. "A poly(A) ribonuclease controls the cellotriose-based interaction between *Piriformospora indica* and its host arabidopsis." *Plant Physiology* 176 (3): 2496–2514.

Jones, K. M., H. Kobayashi, B. W. Davies, M. E. Taga, and G. C. Walker. 2007. "How rhizobial symbionts invade plants: the *Sinorhizobium-Medicago* model." *Nature Reviews Microbiology* 5 (8): 619–633.

Keswani, C., Mishra S., Sarma B. K., Singh S. P., and H. B. Singh. 2014. "Unraveling the efficient applications of secondary metabolites of various *Trichoderma* spp." *Applied Microbiology Biotechnology* 98: 533–544. 10.1007/s00253-013-5344-

Khan, W., B. Prithiviraj, and D. L. Smith. 2008. "Nodfactor[NodBj V(C18:1,MeFuc)] and lumichrome enhance photosynthesis and growth of corn and soybean." *Journal of Plant Physiology* 165 (13): 1342–1351.

Klein, D., and D. E. Eveleigh. 1998. "Ecology of *Trichoderma*." In *Trichoderma*, edited by C. Gliocladium, P. Kubocek, and G. E. Harman, 1: 57–74. London, UK: Taylor & Francis.

Kong, W., Meldgin D. R., Collins J. J., and T. Lu. 2018. "Designing microbial consortia with defined social interactions." *Nature Chemical Biology* 14: 821–829. 10.1038/s41589-018-0091-7

Leff, J. W., Lynch R. C., Kane N. C., and N. Fierer. 2016. "Plant domestication and the assembly of bacterial and fungal communities associated with strains of the common sunflower, *Helianthus annuus*." *New Phytology* 214: 412–423. 10.1111/nph.14323

Lombardi, N., Vitale S., Turrà D., Reverberi M., Fanelli C., and F. Vinale, et al. 2018. "Root exudates of stressed plants stimulate and attract *Trichoderma* soil fungi." *Molecular Plant-Microbe Interactions* 31: 982–994. 10.1094/MPMI-12-17-0310-R

Long, B. M., W. Y. Hee, and R. E. Sharwoodetal. 2018. "Carboxysome encapsulation of the CO2-fixing enzyme Rubisco in tobacco chloroplasts." *Nature Communications* 9 (1): 3570.

Lorito, M., and S. L. Woo. 2015. "*Trichoderma*: a multi-purpose tool for integrated pest management." In *Principles of Plant-Microbe Interactions*. Edited by B. Lugtenberg, 345–353 Cham: Springer International Publishing. 10.1007/978-3-319-08575-3_36

Lorito, M., S. L. Woo, G. E. Harman, and E. Monte. 2010. "Translational research on *Trichoderma*: from "omics" to the field." *Annual Review of Phytopathology* 48 (1): 395–417.

Lugtenberg, B. 2015. *Principles of Plant-Microbe Interactions: Microbes for Sustainable Agriculture*. Cham: Springer International Publishing, p. 448.

Mallock, D. 2018. "Natural history of fungi/mycorrhizae." http://nbm-mnb.ca/mycologyweb-pages/NaturalHistoryOfFungi/ Mycorrhizae.html.

Manganiello, G., Sacco A., Ercolano M. R., Vinale F., Lanzuise S., and A. Pascale, et al. 2018. "Modulation of tomato response to *Rhizoctonia solani* by *Trichoderma harzianum* and its secondary metabolite harzianic acid." *Frontier Microbiology* 9: 1966. 10.3389/fmicb.2018.01966

Markmann, K., and M. Parniske. 2009. "Evolution of root endosymbiosis with bacteria: how novel are nodules?" *Trends in Plant Science* 14 (2): 77–86.

Martínez-Medina, A., Van Wees S., and C. M. Pieterse. 2017. "Airborne signals by *Trichoderma* fungi stimulate iron uptake responses in roots resulting in priming of jasmonic acid-dependent defences in shoots of *Arabidopsis thaliana* and *Solanum lycopersicum*." *Plant Cell Environment* 40: 2691–2705. 10.1111/pce.13016

Mastouri, F., Björkman T., and G. E. Harman. 2012. "*Trichoderma harzianum* enhances antioxidant defense of tomato seedlings and resistance to water defecit." *Molecular Plant-Microbe Interactions* 9: 1264–1271. 10.1094/MPMI-09-11-0240

Morán-Diez, E., Hermosa R., Ambrosino P., Cardoza R. E., Gutiérrez S., and M. Lorito, et al. 2009. "The ThPG1 endopolygalacturonase is required for the *Trichoderma harzianum*–plant beneficial interaction." *Molecular Plant-Microbe Interactions* 22: 1021–1031. 10.1094/MPMI-22-8-1021

Neumann, B. and M. Laing, 2006. "Trichoderma: an ally in the quest for soil system sustainability." In *Biological Approaches to Sustainable Soil Systems*. Edited by N. Uphoff et al, 491–500. Boca Raton, FL: CRC Press.

O'Malley, M. A. 2017. "From endosymbiosis to holobionts: evaluating a conceptual legacy." *Journal of Theoretical Biology* 434: 34–41.

Ort, D. R., S. S. Merchant, J. Alric, A. Barkan, R. E. Blankenship, R. Bock, R. Croce, M. R. Hanson, J. M. Hibberd, S. P. Long, T. A. Moore, J. Moroney, K. K. Niyogi, M. A. J. Parry, P. P. Peralta-Yahya, R. C. Prince, K. E. Redding, M. H. Spalding, K. J.

van Wijk, W. F. J. Vermaas, S. von Caemmerer, A. P. M. Weber, T. O. Yeates, J. S. Yuan, and X. Zhu. 2015. "Redesigning photosynthesis to sustainably meet global food and bioenergy demand." *Proceedings of the National Academy of Sciences* 112 (28): 8529–8536.

Panstruga, R., and H. Kuhn. 2015. "Introduction to a Virtual Special Issue on cell biology at the plant-microbe interface." *New Phytologist* 207 (4): 931–938.

Parniske, M. 2008. "Arbuscular mycorrhiza: the mother of plant root endosymbioses." *Nature Reviews Microbiology* 6 (10): 763–775.

Perez-Jaramillo, J. E., Mendes R., and J. M. Raaijmakers. 2016. "Impact of plant domestication on rhizosphere microbiome assembly and functions." *Plant Molecular Biology* 90: 635–644. 10.1007/s11103-015-0337-7

Philippot, L., Raaijmakers J. M., Lemanceau P., and W. H. van der Putten. 2013. "Going back to the roots: the microbial ecology of the rhizosphere." *Nature Review Microbiology* 11: 789–799. 10.1038/nrmicro3109

Porcel, R., R. Aroca, and J. M. Ruiz-Lozano. 2011. "Salinity stress alleviation using arbuscular mycorrhizal fungi. A review." *Agronomy for Sustainable Development* 32 (1): 181–200.

Porcel, R., S. Redondo-Go´mez, E. Mateos-Naranjo, R. Aroca, R. Garcia, and J. M. Ruiz-Lozano. 2015. "Arbuscular mycorrhizal symbiosis ameliorates the optimum quantum yield of photosystem II and reduces non-photochemical quenching in rice plants subjected to salt stress." *Journal of Plant Physiology* 185: 75–83.

Rizvi, A., and M. S. Khan. 2018. "Heavy metal induced oxidative damage and root morphology alterations of maize (*Zea mays* L.) plants and stress mitigation by metal tolerant nitrogen fixing *Azotobacter chroococcum*." *Ecotoxicology Environtal Safety* 157: 9–20. 10.1016/j.ecoenv.2018.03.063

Rojas-Tapias, D., Moreno-Galván A., Pardo-Díaz S., Obando M., Rivera D., and R. Bonilla. 2012. "Effect of inoculation with plant growth-promoting bacteria (PGPB) on amelioration of saline stress in maize (*Zea mays*)." *Applied Soil Ecology* 61: 264–272. 10.1016/j.apsoil.2012.01.006

Rubio, B. M., Quijada N. M., Pérez E., Domínguez S., Monte E., and R. Hermosa. 2014. "Identifying beneficial qualities of *Trichoderma parareesei* for plants." *Applied Environmental Microbiology* 6: 1864–1873. 10.1128/AEM.03375-13

Ruocco, M., S. Lanzuise, N. Lombardi, S. L. Woo, F. Vinale, R. Marra, R. Varlese, G. Manganiello, A. Pascale, V. Scala, D. Turrà, F. Scala, M. Loritoet. 2015. "Multiple roles and effectsof a novel *Trichoderma* hydrophobin." *Molecular Plant-Microbe Interactions* 28 (2): 167–179.

Ruzzi, M., and R. Aroca. 2015. "Plant growth-promoting rhizobacteria act as biostimulants in horticulture." *Science Horticulturae* 196: 124–134. 10.1016/j.scienta.2015.08.042

Samolski, I., Rincón A. M., Pinzón L. M., Viterbo A., and E. Monte. 2012. "The qid74 gene from *Trichoderma harzianum* has a role in root architecture and plant biofertilization." *Microbiology* 158: 129–138. 10.1099/mic.0.053140-0

Samuels, G. J., and P. K. Hebbar. 2015. "Identification and agricultural properties." *The American Phytopathological Society* 32. St. Paul, MN: The American Phytopathological Society.

Schmidt, D., E. K. James, G. J. Ellis, J. E. Shaw, and J. I. Sprent. 1994. "Interaction between rhizobia and potato tissues." *Journal of Experimental Botany* 45 (279): 1475–1482.

Shi, S., Nuccio E. E., Shi Z. J., He Z., Zhou J., and M. K. Firestone. 2016. "The interconnected rhizosphere: high network complexity dominates rhizosphere assemblages." *Ecology Letter* 19: 926–936. 10.1111/ele.12630

Shoresh, M., and G. E. Harman. 2008. "The molecular basis of shoot responses of maize seedlings to *Trichoderma harzianum* T22 inoculation of the root: a proteomic approach." *Plant Physiology* 147 (4): 2147–2163.

Shoresh, M., Harman G. E., and F. Mastouri. 2010. "Induced systemic resistance and plant responses to fungal biocontrol agents." *Annual Review Phytopathology* 48: 21–43. 10.1146/annurev-phyto-073009-114450

Sivasithamparam, K., and E. L. Ghisalberti. 1998. "Secondary metabolism in *Trichoderma* and *Gliocladium*." In *Trichoderma and Gliocladium. Volume 1: Basic Biology, Taxonomy and Genetics*. Edited by G. E. Harman, and C. P. Kubicek, 139–191. London: Taylor and Francis Ltd.

Smith, R. S., and H. Absan. 2013. A. chito-ologiosaccharides and methods for use in enhancing corn growth. US patent application 2013/0079225.

South, P. F., A. P. Cavanagh, H. W. Liu, and D. R. Ort. 2019. "Synthetic glycolate metabolism pathways stimulate crop growth and productivity in the field." *Science* 363 (6422): eaat9077.

Strack, D. T. Fester, B. Hause, W. Schliemann, and M. H. Walter. 2003. "*Arbuscular mycorrhiza*: biological, chemical, and molecular aspects." *Journal of Chemical Ecology* 29 (9): 1955–1979.

Stringlis, I. A., Zhang H., Pieterse C. M. J., Bolton M. D., and R. de Jonge. 2018. "Microbial small molecules - weapons of plant subversion." *Nature Production Reports* 35: 410–433. 10.1039/c7np00062f

Studholme, D. J., Harris B., Le Cocq K., Winsbury R., Perera V., and L. Ryder, et al. 2013. "Investigating the beneficial traits of *Trichoderma hamatum* GD12 for sustainable agriculture – insights from genomics." *Frontier Plant Science* 4: 258. 10.3389/fpls.2013.00258

Tahat, M. M., N. Nejat, and K. Sijam. 2014. "*Glomus mosseae* bioprotection against aster yellows phytoplasma (16srI-B) and *Spiroplasma citri* infection in Madagascar periwinkle." *Physiological and Molecular Plant Pathology* (88): 1–9.

Vacheron, J., Desbrosse G., Bouffaud M. L., Touraine B., Moënne-Loccoz Y., and D. Muller, et al. 2013. "Plant growth-promoting rhizobacteria and root system functioning." *Frontier Plant Science* 4: 356. 10.3389/fpls.2013.00356

Van Oosten, M. J., Di Stasio E., Cirillo V., Silletti S., Ventorino V., and O. Pepe, et al. 2018. "Root inoculation with *Azotobacter chroococcum* 76A enhances tomato plants adaptation to salt stress under low N conditions." *BMC Plant Biology* 18: 205. 10.1186/s12870-018-1411-5

Van Oosten, M. J., Pepe O., De Pascale S., Silletti S., and A. Maggio. 2017. "The role of biostimulants and bioeffectors as alleviators of abiotic stress in crop plants." *Chemical and Biological Technologies in Agriculture* 4: 5. 10.1186/s40538-017-0089-5

Ventorino, V., Pascale A., Adamo P., Rocco C., Fiorentino N., and M. Mori, et al. 2018. "Comparative assessment of autochthonous bacterial and fungal communities and microbial biomarkers of polluted agricultural soils of the Terra dei Fuochi." *Science Report* 8: 14281. 10.1038/s41598-018-32688-5

Ventorino, V., Sannino S., Piccolo A., Cafaro V., Carotenuto R., and O. Pepe. 2014. "*Methylobacterium populi* VP2: plant growth-promoting bacterium isolated from a highly polluted environment for polycyclic aromatic hydrocarbon (PAH) biodegradation." *Science World Journal* 2014: 1–11. 10.1155/2014/931793

Vinale, F., Flematti G., Sivasithamparam K., Lorito M., Marra R., and B. W. Skelton, et al. 2009. "Harzianic acid, an antifungal and plant growth promoting metabolite from *Trichoderma harzianum*." *Journal of Natural Products* 72: 2032–2035. 10.1021/np900548p

Vinale, F., M. Nigro, K. Sivasithamparam, G. Flematti, E. L. Ghisalberti, M. Ruocco, R. Varlese, R. Marra, S. Lanzuise, A. Eid, S. L. Woo, and M. Lorito. 2013. "Harzianic acid: a novel siderophore from *Trichoderma harzianum*." *FEMS Microbiology Letters* 347 (2): 123–129.

Vinale, F., Manganiello G., Nigro M., Mazzei P., Piccolo A., and A. Pascale, et al. 2014. "A novel fungal metabolite with beneficial properties for agricultural applications." *Molecules* 19: 9760–9772. 10.3390/molecules19079760

Vinale, F., Sivasithamparam K., Ghisalberti E. L., Marra R., Barbetti M. J., and H. Li, et al. 2008. "A novel role for *Trichoderma* secondary metabolites in the interactions with plants." *Physiological and Molecular Plant Pathology* 72: 80–86. 10.1016/j.pmpp.2008.05.005

Viscardi, S., Ventorino V., Duran P., Maggio A., De Pascale S., and M. L. Mora, et al. 2016. "Assessment of plant growth promoting activities and abiotic stress tolerance of *Azotobacter chroococcum* strains for a potential use in sustainable agriculture." *Journal of Soil Science and Plant Nutrition* 16: 848–863. 10.4067/S0718-95162016005000060

Wani, S. A., Chand S., and T. Ali. 2013. "Potential use of *Azotobacter chroococcum* in crop production: an overview." *Current Agricultural Research Journal* 1: 35–38. 10.12944/CARJ.1.1.04

Wargo, M. J., and D. A. Hogan. 2006. "Fungal-bacterial interactions: a mixed bag of mingling microbes." *Current Opinion Microbiology* 9: 359–364. 10.1016/j.mib.2006.06.001

*Trichoderma lignorum* on *Rhizoctonia solani* and other soil fungi." *Phytopathology* 24: 1153–1179.

Weindling, R. 1932. "*Trichoderma lignorum* as a parasite of other soil fungi." *Phytopathology* 22: 837–845.

Woo, S. L., Ruocco M., Vinale F., Nigro M., Marra R., and N. Lombardi, et al. 2014. "*Trichoderma*-based products and their widespread use in agriculture." *Open Mycology J.* 8: 71–126. 10.2174/1874437001408010071

Wu, Q., X. Peng, M Yang, W. Zhang, F. B. Dazzo, N. Uphoff, Y. Jing, and S. Shen. 2018. "Rhizobia promote the growth of rice shoots by targeting cell signaling, division and expansion." *Plant Molecular Biology* 97 (6): 507–523.

Yakhin, O. I., Lubyanov A. A., Yakhin I. A., and P. H. Brown. 2017. "Biostimulants in plant science: a global perspective." *Frontier Plant Science* 7: 2049. 10.3389/fpls.2016.02049

Yanni, Y. G., F. B. Dazzo, A. Squartini, M. Zanardo, M. I. Zidan, and A. E. Y. Elsadany. 2016. "Assessment of the natural endophytic association between *Rhizobium* and wheat and its ability to increase wheat production in the Nile delta." *Plant Soil* 407 (1-2): 367–383.

Yanni, Y. G., F. B. Dazzo, and M. I. Zidan. 2011. "Beneficial endophytic Rhizobia as biofertilizer inoculants for rice and the spatial ecology of this bacteria-plant association." In *Bacteria in Agrobiology: Crop Ecosystems.* Edited by D. K. Maheshwari, 265–294.

Yanni, Y. G., R. Y. Rizk, F. K. A. El-Fattah, A. Sqartini, V. Corich, A. Giacomini, F. J. de Bruijn, J. L.W. Rademaker, J. Maya-Flores, P. Ostrom, M. Vega-Hernandez, R. I. Hollingsworth, E. Martinez-Molina, P. Materos, E. Velazquez, J. Wopereis, E. Triplett, M. U. Garcia, J. A. Anarna, B. G. Rolfe, J. K. Ladha, J. E. Hill, R. Mujoo, P. K., Ng, and F. B. Dazzo. 2001. "Beneficial plant growth-promoting association of *Rhizobium leguminosarum* bv. trifolii with rice roots." *Functional Plant Biology* 28 (9): 845–870.

Yanni, Y. G., R. Y. Rizk, V. Corich, A. Squartini, K. Ninke, S. Philip-Hollingsworth, G. Orgambide, F. de Bruijn, J. Stoltzfus, D. Buckley, T.M. Schmidt, P.F. Mateos, J.K. Ladha, and F. B. Dazzo. 1997. "Natural endophytic association between *Rhizobium leguminosarum* bv. trifolii and rice roots and assessment of its potential to promote rice growth." *Plant and Soil* 194 (1/2): 99–114.

Yanni, Y., M. Zidan, F. Dazzo, R. Rizk, A. Mehesen, F. Abdeldattah, and A. Elsadany. 2016. "Enhanced symbiotic performance and productivity of drought stressed common bean after inoculation with tolerant native rhizobia in extensive fields." *Agriculture, Ecosystems and Environment* 232: 119–128.

Zachow, C., C. Berg, H. Müller, J. Monk, and G. Berg. 2016. "Endemic plants harbour specific Trichoderma communities with an exceptional potential for biocontrol of phytopathogens." *Journal of Biotechnology* 235: 162–170.

# 3 Dynamic Role of Specific Microbes in Bioremediation of Heavy Metals and Dyes from the Textile Industry

*Tamanna Sultana, Debapriya Maitra,*
*Bedaprana Roy, Arup Kumar Mitra,*
*and Xavier Savarimuthu, SJ*

## CONTENTS

## 3.1 INTRODUCTION – TEXTILE AND DYEING INDUSTRIES

The increased rate of urbanization and industrialization has led to an enormous amount of wastes being generated and deposited, resulting in environmental pollution and other detrimental consequences (Senan and Abraham 2004). Synthetic dyes are used on a large-scale basis in factories and industrial locations associated with pharmaceutical, food, and cosmetic production, and these dyes are also widely used in textile industries and photography locations of certain industrial sectors. There are different types and classes of dyes available, depending on which textile industries discharge variable fractions of dyes as effluents to the environment. Industries that use basic dye only have a dye effluent loss of around 1–2%, while those using reactive dyes have an effluent loss of 50% or more. Approximately 22% of the total effluent loss enters the environment from wastewater treatment plants (Kirk and Othmer 1993; Sheth and Dave 2009). One such group of dyes that exert hazardous effects on the environment are azo dyes. Azo dyes have toxic consequences on health and the environment due to the reactive nature of the color components present in

31

32                                           Go Green for Environmental Sustainability

the dye. Biocalcitrance and toxic effects of azo dyes on humans and animals are concerning issues found in this dye category and are presently being widely studied to combat the problem and to provide an effective solution. The conventional and traditional physical and chemical methods used to treat the waste and effluents generated from textile and dyeing industries are mostly rendered ineffective due to the extreme properties of these effluents, such as high biochemical oxygen demand (BOD), high chemical oxygen demand (COD), and presence of metals, which are often heavy metals. Practically, these treatment methods are highly expensive, generate a large quantity of sludge, and produce harmful and undesirable byproducts and toxic substances (Senan and Abraham 2004). A matter of concern in this discussion is that there is presently no evidence of enzymes secreted by the human digestive system that act upon azo dyes. Subsequently, the absence of an enzyme-mediated degradation ability of human metabolic systems leads to the production and accumulation of carcinogenic and toxic substances in the human body. Thus, potable water contaminated with azo dyes should be taken care of with the utmost importance (Kirk and Othmer 1993; Sheth and Dave 2009).

Procedures and techniques used for dye treatment are divided into three broad categories: (1) physical methods, (2) chemical methods, and (3) biological methods. The prevailing physical and chemical methods include surface absorption or adsorption, coagulation and precipitation techniques, radiation, thermal degradation or incineration, and ultrafiltration. These procedures do have negative aspects:

1. These methods and associated instruments and apparatus are difficult to handle using cumbersome operational requirements.
2. High expenditure and capital investment are required for setting up these units and for maintenance costs (Georgiou et al. 2003).

India's textile and dyeing industry plays a pivotal role in the country's economic development through creating clothing materials, which are a basic necessity. This industry provides a major contribution to the gross industrial output, generating employment opportunities and supporting the national economy through export of goods. Indian textile industry contributes around 4% of the country's total GDP, 10% to the country's manufacturing stakes, and is regarded as the country's second largest employment generator. The industry is predicted to have a compound annual growth rate (CAGR) of 3.7% during the financial span of 2018–2028 [Equitymaster, 2020].

Certain dyes are detrimental to humans and the environment because some dyes do not adhere to the fabric. During downstream processing, these dyes are deposited as wastewater in soil and bodies of water in the surrounding or adjacent areas of the factory and textile dyeing locations. The use of various dyes in textile industries is increasing across the country. These dyes bind to surfaces of fabric or other materials by adsorption, by forming chemical covalent bonds, or by forming complexes with metals or salts, both in solid mixtures and in solution. The ability of specific chemical groups or components to change the color of surfaces makes them useful and widely demanded as dye components in various industries. The particular color of a dye depends upon its ability to absorb light of a specific wavelength within the visible spectrum. Dyes are popularly used in many industries, such as textiles, cosmetics, food and pharmaceuticals, paper printing, and others. Historically, dyes were mostly derived from natural sources, such as animals or the flowering parts of plants. Interestingly, microbes convert the organic components of dyes to inorganic forms through their own metabolic processes and decomposition procedures. Synthetic dyes now dominate the market and industrial sectors. Unfortunately, the microbial populations in nature are unable to degrade the synthetic dyes. Additionally, each synthetic dye has specific and unique physical and chemical properties that hinder predicting the full impact on human health and the environment. The cumulative impact of dye wastewater and the increase in the use of synthetic dyes poses a serious threat, resulting in textile dyeing industries becoming a key source of environmental pollution. Though the toxicity level of these dyes is generally low, there are other inherent factors and properties through which synthetic dyes cause serious health and environmental hazards, even at low concentrations. At concentrations of only 1 ppm, synthetic dyes can cause turbidity in water, indicating the potential of these dyes in contaminating bodies of water even at low concentrations. Additionally, textile dyeing industries that discharge synthetic dyes as effluent have been observed to cause reduction in light penetration ability, increased BOD of water, and decreased dissolved oxygen content in surrounding water bodies (Chen et al. 2003). The prevailing environmental implications and public health concerns existing locally around the dye facilities and industries demand better alternative treatment procedures and proper bioremediation of polluted sites. Owing to the possible negative impacts of synthetic dye on public health and on the ecosystem, strict regulations regarding dye effluent treatment must be executed. Microorganisms like white rot fungi, bacteria, and their consortia can effectively decolorize a wide variety of dyes (Asghar et al. 2015), and researchers are exploiting this unique capability to remove dye residues from the environment in a sustainable and cost-effective manner (Verma and Madamwar 2003; Moosvi et al. 2005).

## 3.2  USE OF A BACTERIAL CONSORTIUM FOR EFFECTIVE BIOREMEDIATION

Textile industries are known for being one of the largest global consumers of freshwater and chemicals, while also being one of the largest contributors to global pollution. The major heavy metals found in the textile industry's dye effluents mainly consist of cadmium, lead, and zinc (Prasad et al. 2006; Hill et al. 1993), along with residues of reactive metal complexes from chromophore dyes, like copper, nickel, cobalt, chromium, etc. (Sungur and Gülmez 2015). Although there are a number of methods that can be used to mitigate the toxicity of textile effluents, one of the most effective and environmentally friendly techniques involves the use of indigenous microflora from soil. Indigenous microflora present in soil mainly belonging to the *Bacillus* family and *Pseudomonas* family are known to possess several properties for bio-sorption of heavy metals from industry effluents (Ajao et al. 2011). Certain studies have shown that various strains of *Escherichia coli* adsorb heavy metal at higher levels than other bacterial species (Jin et al. 2018). In addition to bacterial species, various fungal species, especially arbuscular mycorrhizal (AM) fungi and

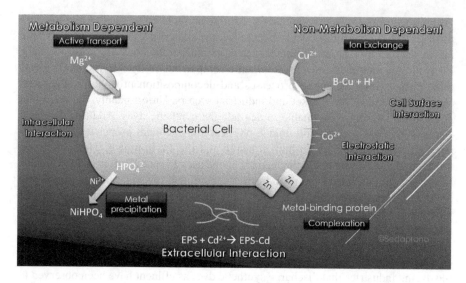

**FIGURE 3.1** Bio-absorbance of metals by microbes.

fungi belonging to *Penicillium* sp., *Rhizopus* sp., and *Aspergillus* sp. (Huang and Huang. 1996), have shown potential for heavy metal remediation properties (Volesky and Holan 1995; Huang et al. 1985–1990). Some algal species, like *Ulva fasciata and Sargassum densifolium* (El-Wakeel et al. 2019), also demonstrate promising bio-absorbance on heavy metals, like copper(II) ion concentrations. The associated mechanism has been depicted in Figure 3.1.

The use of microorganisms to update or remediate heavy metals can occur through various means. Normally, the various functional groups present on the microbial cell surface, including the carboxyl group, hydroxyl group, phosphate group, amino group, etc., bind with heavy metals (Scott and Karanjkar 1992). This binding and uptake can be an active process or a passive process (Hussain et al. 2001). A feasible means of this bioremediation is through the adsorption process. According to a study by Wang et al. (2001), adsorption is the primary non-energy driven method for heavy metal uptake in microbes. Reports from He and Tebo (1998) show that species of genus *Bacillus* found at pH 7.2 can adsorb as much as up to 60% of soil copper(II) ions within only 1 minute, reaching the adsorption equilibrium within 10 minutes. Conversely, the active energy process of bio-absorption has been found to be comparatively less efficient. A third common method for bioremediation is bio-mining, where the heavy metal ions from the ores are mobilized or oxidized by forming bio-complexes (Brunetti et al. 2012; Volesky and Holan 1995) or through bio-oxidation (Rahman and Sathasivam 2015).

The microorganisms' potential to uptake or leach heavy metals varies from species to species and depends on a number of factors. For example, the leaching rate of microbes under nutrient-unavailable condition is around 9%; however, with the addition of glucose and other nutrients, this leaching rate increases to 36% (Chanmugathas and Bollag 1988). The dead or passive cells of *Bacillus licheniformis* RO8 are effective in reducing lead(II) ions (Pb$^{2+}$) to the PbO, lead(II) oxide, form (Goyal et al. 2003). Different microbes found in separate species have different potentials for removing heavy metals. For example, *Escherichia coli* strain K12 has one of the largest ranges of heavy metal absorption potentials, and its outer cell wall can absorb up to 30 different metals, including Hg, Cd, On, Cu, Ni, and Zn. Some of the spore forming soil bacterial species, such as *Bacillus sp.* and *Pseudomonas sp.*, have higher order Qmax values (maximum adsorption capacity value) than any other bacterial species (Jin et al. 2018). The studies from Gazso (2001), Dermont et al. (2008), Hynninen et al. (2010), Brunetti et al. (2012), and Ehrlich (1997) show the comparative potential of Qmax value of different bacterial species for different heavy metals:

1. *Bacillus subtilis* has a Qmax value of 101 for cadmium and a value of 137 for zinc.
2. *Pseudomonas aeruginosa* has a reduced Qmax value of 57.37 for cadmium, a value of 13.7 for zinc, and a value of 79.5 for lead. The mechanism of cadmium bioremediation has been depicted in Figure 3.2.
3. *Streptomyces noursei* has a low Qmax potential of 3.4 for cadmium, 1.6 for zinc, and 99 for lead, respectively.
4. *Bacillus licheniformis* has a high adsorption capacity for metals, such as 142.7 for lead and 62 for chromium.
5. *Bacillus laterosporus* has Qmax value of 159.2 for cadmium and 72.6 for chromium.

The commonly used beneficial PGPR fungi *Rhizopus arrhizus* has Qmax values of 26.8 for cadmium, 4.5 for chromium, and 55.6 for zinc. According to a study by Franke et al. (2003), there are three main types of efflux transporters for conferring

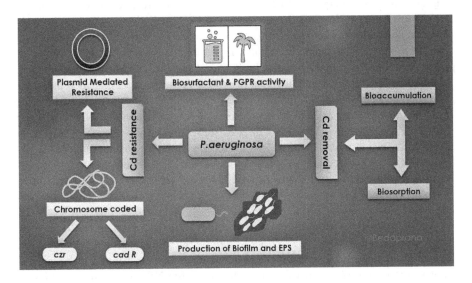

**FIGURE 3.2** Schematic representation of cadmium bioremediation.

heavy metal resistance to bacteria: (1) CBA type, (2) P-type ATPase, and the (3) CDF (cationic diffusion facilitator) family. The rare and exceptional CBA type has a tri-component transmembrane arrangement with a RND protein (resistance, nodulation, and cell division), a membrane fusion protein (MFP), and an outer membrane protein (OMP). The other two common transporters found in different bacteria are the P-type ATPase transporter, which pumps ionic efflux from cytoplasm toward periplasm, and the CDF family of transporters (Anton et al. 1999; Grass et al. 2001). Nies in his study (2003) reported that the presence of CBA type transporter with RND protein indicates a very high level of metal resistance. Generally, these efflux pumping systems have both general resistance for similar types of metals, like $Zn^{2+}$, $Co^{2+}$, $Cu^{2+}$, and $Pb^{2+}$ ions, etc., and specific resistance genes, like pbrTRABCD, for lead transport (Choudhary et al. 2017). Examples of a general transport pump can be the zntA gene present in *Escherichia coli* and the cadA2 frame from *Pseudomonas putida* for effective transport of lead, zinc, and cadmium (Chellajah et al. 2018; Marchler-Bauer et al. 2007). The czcCBA1 gene of CBA family transporter in *Pseudomonas putida* KT2440 can be an example of a specific transporter pump as it only takes part in $Pb^{2+}$ export of cell (Hynninen et al. 2010).

A comprehensive mitigation of heavy metals from the textile industry's effluents can be attempted through using a combination of microbes. Several studies have demonstrated that the efficiency of heavy metal removal and azo-dye discoloration is higher in consortium-based live cultures and immobilized cultures. In a comparative study conducted by Allam (2017), four bacterial isolates, *Sphingomonas paucimobilis, Rhizobium radiobacter, Bacillus subtilis,* and *Bacillus pumilus,* were selected based on their percentage removal efficiency of copper, iron, lead, cadmium, and chromium metals. Both single- and consortium-based cultures were tested, and the consortium form, for both immobilized and free cell cultures, had better efficiency. These results agree with the findings of Khehra et al. (2005) and Ali (2010), who suggested that the synergistic effects of the metabolism of different microbes used in the consortium provides an enhanced scope of bioremediation, compared to single bacterial cell cultures. As these soil bacteria are loaded with plant growth-promoting activities, using these bacterial isolates in combination for the holistic improvement of various crop plants has also been widely explored. In a study attempted by Mahmood et al. (2013), three different microbial consortia were designed using six different bacteria – *Bacillus subtilis, Bacillus cereus, Bacillus mycoides, Bacillus* sp., *Pseudomonas* sp., and *Micrococcus* sp. – chosen based on their bioremediation ability (Cu, Cd, Cr, Ni, Mn, and Pb). The study showed percentage reduction in Cu, Cd, Cr, Ni, Mn, and Pb at 92.3%, 89.46%, 83.52%, 80.7%, 88.3%, and 93.5%, respectively. The subsequent treatment of the remediated effluents using living consortium was tested on seedlings of *Zea mays* (maize), which showed an approximate germination taste as high as 94% with significant plant growth rate with respect to the control setup.

A comprehensive microbial consortium composed of various PGPR isolates, like *Bacillus subtilis, Bacillus thuringiensis, Bacillus licheniformis,* and *Pseudomonas* sp., with other rhizospheric isolates has been found to be significant in increasing plant growth rate, alleviating the NPK ratio, and in enhancing the induced systemic resistance response pathway in plants, along with bio-sorbing the percentage heavy metal in soil. In vitro tests of such formulations over various major pathogens of vegetable crops, like cucumber and potato, and on pathogens of cash crops, like tea (*Camellia sinensis*), have been conducted with positive responses. There is further elaborative study needed in unravelling the mechanisms and kinetics of heavy metal bioremediation through microbial consortia, testing, and exploring on both consortia and remediated effluents on various crop plants.

## 3.3   ROLE OF EXTREMOPHILES IN BIOREMEDIATION

Extremophiles are organisms that can survive at the extremes of environmental conditions:

1. Temperature extremes, from psychrophiles to thermophiles to hyperthermophiles
2. pH extremes, from acidophiles to alkalophiles
3. Radiation extremes
4. Salinity extremes, such as halophiles
5. Pressure extremes, such as barophiles

Extremophiles can also tolerate heavy metal concentrations or a combination of these varied conditions. These microbial strains are usually very robust and tend to secrete stable enzymes called extremozymes, as briefly summarized in Figure 3.3,

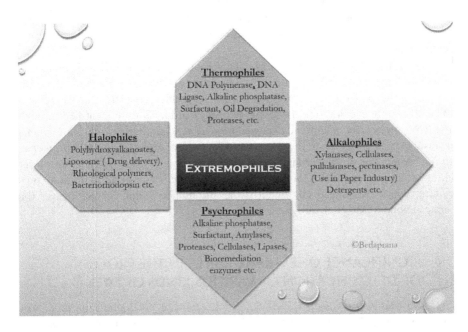

**FIGURE 3.3**   Diverse range of enzymes by extremophilic microorganisms.

that carry out their catalytic activity even at conditions that render the effectiveness of typical bioconversions unviable (Peeples 2014).

Textile industry effluents contain a number of heavy metals (such as chromium, iron, zinc, lead, copper, and manganese), salts, and dyes (usually a nitrophenol derivative and sulphide or polysulfide) (Hussain et al. 2004). These effluents also contain bacteria that are salt tolerant or acid tolerant, can survive in these effluents with altered chemical compositions and metabolically adapt to them, and convert the pollutants into less harmful substances (Peeples 2014).

As previously mentioned, textile effluents contain Azo dyes, which is the largest class of chemical dyes that give large variations in color and is most widely used. There are other classes of dyes: (1) Anthraquinone dyes (neutral, cationic, or anionic); (2) indigoid dyes, which are scarcely soluble due to the presence of inter and intra molecular hydrogen bonding; and (3) metal complex dyes, which contain a heavy metal atom (Co, Cr, Cu) that provides for deep coloration and aids fixation, but also contributes to the heavy metal contamination of textile industry wastewaters (De Oliveira et al. 2018; Said 2018).

The problem is that only 10–15% of the dye is contained in the fabric while the rest is released into the wastewaters as a colored effluent. When this colored effluent is released into the environment, it diminishes the photosynthesis of phytoplankton by decreasing light penetration into lower layers of water bodies, and it deteriorates the water quality (Chen et al. 2003). These dyes are not readily degraded; therefore, they persist in the environment for long periods, and are not only toxic, but also mutagenic and carcinogenic (Zhang et al. 2004).

There are a number of enzymes produced by bacteria that are capable of degrading textile dyes, including ligninolytic peroxidases (lignin peroxidase, horseradish peroxidase, manganese peroxidase), laccases, and azoreductases. These enzymes have been reported to cause dye degradation in textile effluents (Chatha et al. 2017; Sarkar et al. 2017).

Halophilic and halo tolerant strains of bacteria isolated from the textile industry's effluents were found to have the decolorizing ability because they could use the components of the dyes as their sole carbon source. The bacterial isolates that were found to be efficiently decolorizing the textile industry's effluents were identified through 16srRNA sequencing and mainly showed phylogenetic similarity to the *Halomonas* genus, such as *Halomonas aquamarina*, *Halomonas meridiana*, and *Halomonas salina*. A number of further analytical studies, such as UV-Vis analysis, thin-layer chromatography, and HPLC analysis, showed that the bacterial decolorization procedure reduces the azo bond, which is the chromophore, and further cleaves the reduced bond (Asad et al. 2007).

## 3.4 SIGNIFICANCE OF EXTREMOPHILES IN THE PROCESS OF DETOXIFICATION OF VARIOUS KINDS OF DYES

*Nesterenkonia lacusekhoensis* EMLA3, a halo tolerant alkalophilic bacteria, was found to degrade Methyl Red under alkaline conditions. This bacterium was isolated from alkaline textile effluent. Rapid decolorization of the dyes were observed both in

NB and MSM, along with no decolorization of uninoculated medium. The GC-MS study of the degraded components showed benzeneacetamide and acetophenone, and FTIR study showed presence of *N,N'*dimethyl-*p*-phenyle-nediamine (DMPD) or its derivatives (Jadhav et al. 2008; Gomare and Govindwar 2009) and aminobenzoic acid by azo bond cleavage (Zhao et al. 2014). *Nesterenkonia lacusekhoensis* EMLA3 decolorized 97% of 50 mg/L Methyl Red after 16 hours of incubation, though the decolorization efficiency of the strain was found to decrease with increasing concentrations of Methyl Red. This may be due to higher concentrations of Methyl Red conferring some kind of toxicity of the bacterium, and increasing the inoculum size was found to solve the problem. The strain degraded methyl red at a wide range of alkaline pH (8–11.5), a characteristic that is desirable because the textile industry wastewaters tend to have this alkaline range of pH. The strain EMLA3 showed appreciable activity of dye removal in real textile effluent containing heavy metals, salinity, and high alkalinity, with even a reduction in COD) (Bhattacharya et al. 2017).

AMETH 156, another halo tolerant marine bacterium strain, was filtered out from 30 bacterial isolates from the waters of the Kelambakkam saltpans. These bacterial isolates were filtered based on their ability to degrade several dyes present in the textile industry's effluents, such as Magenta P, Black B, and Yellow MR. They were also screened on the basis of indole acetic acid IAA and auxine production; *Halomonas* sp. was found to be the best. It was then immobilized with calcium alginate, causing considerable improvement in dye decolorization ability (Sumathi et al. 2018). Bio-augmentation and bio-stimulation strategies have been found to boost the process of dye degradation and decolorization through adding nutrients to the bacterial consortium (Lade et al. 2015).

Extremophillic bacteria's unique molecular mechanisms make them capable of acting as micro factories for metal bioremediation. Heavy metals like chromium, zinc, iron, copper, etc. also contribute to the textile industry's effluent's composition, and remediation of these are necessary, as metal pollution leaves a huge environmental hazard footprint (Marques et al. 2014). Their persistence in the environment causes bioaccumulation in the trophic chain and exposes the living organisms to these xenobiotics (Hussein et al. 2001, Douay et al. 2013).

The extremophilic organisms' membrane structure contributes to their ability to quickly adapt in their transcriptional activity, and their translational activity allows them to activate or inhibit metal binding, anti-oxidative stress mechanisms, and metal permeability responses (Dekker et al. 2016; Singh and Singh 2017). The extremophiles' membrane structures are characterized with a positively charged inner layer and metal transporter proteins (Kumar et al. 2018), which control metal and protein entry to fight acid stress, metal stress, and radionuclide stress (Navarro et al. 2013; Zhang et al. 2016).

Metal biosorption by extremophiles under stressed conditions is helped by their cell wall/capsule, S-layer proteins, and their EPS (extracellular polymer substance), which have certain structural and functional properties that enhance and favor metal biosorption. On the other hand, siderophores (metal scavenging biomolecules produced by extremophiles) have special features and regulatory functions that play a potential role in metal chelation and bioremediation (De Serrano et al. 2016).

Extremophillic bacteria produce extracellular or intracellular enzymes to chemically transform metals, causing redox reactions to occur that directly or indirectly solubilize or precipitate the metal (Das et al. 2016). A number of extremophilic bacteria, such as halophilic (Srivastava et al. 2013), thermophilic, thermotolerant, metallophilic, and acidophilic bacteria, have been reported to precipitate several metals, out of which chromium, iron, copper, and metal sulphides are present in textile industry effluents (Marques 2018).

*Amphibacillus* and *Halomonas* were found to reduce hexavalent chromium up to 70% while under 10% sodium chloride salinity and an alkaline pH of 10 (Ibrahim et al. 2011; Mabrouk et al. 2014). *Bacillus* sp. Vi-1 and *Stenotrophomonas* sp. Vi-4, isolated from the Antarctic soils, were both psychrotolerant and halotolerant. They proved useful in treating water contaminated with heavy metals, such as copper, cadmium, chromium, cobalt, nickel, lead, and zinc at low temperatures (Tomova et al. 2014).

Two halophilic strains of Fungi, isolated from saline conditions in Thailand, were found to be extremophilic in nature: (1) *Aspergillus flavus* and (2) *Sterigmatomyces halophilus*. When these two strains were incubated for 14 days, they showed 85% iron and zinc biosorption capacity (Bano et al. 2018).

Extremophiles are therefore very promising resources, as they contain cellular apparatus and genetic makeup that can easily adapt to harsh environments and help in whole cell bioremediation in such extreme environments for soil and water treatments. These observations call for an in-depth study of extremophiles and polyextremophiles that can survive multiple stressed conditions and produce biomolecules that are active in such stresses, which can help in bioremediation processes. As the area of study of extremophiles remains scarcely explored, there is presently little known about their applications in environmental biotechnology (Giovanella et al. 2019).

## 3.5 CONCLUSION

Approximately 56% of globally annually produced dye is consumed by the textile industry (Adegoke and Bello 2015; Markandeya et al. 2007). The textile industry is also known to be one of the largest consumers of freshwater. The textile industry produces a large discharge of heavy metals, like lead, cadmium, zinc, nickel, cobalt, and chromium, along with a significant amount of effluent derived from various azo-based dyes. With the need for immediate treatment of these textile effluents, the use of various microorganisms for degrading the dyes offers a viable option. There are various filamentous fungi (Siddiquee et al. 2015), algae, and common rhizospheric, hydrospheric, and poly extremophilic bacterial species that have potential metal uptake or dye decolorizing capabilities. Using microorganisms to degrade dyes has both advantages and disadvantages. Bioremediation stands out as one of the most effective and eco-friendly means for effluent and wastewater treatment. However, some of the major drawbacks for bioremediation include the isolation of rapid dye decolorizing microbes from single culture for commercial applications or their abilities to degrade/uptake various kinds of dyes/heavy metals. The use of microorganisms in consortium form or various extremophilic bacteria can be a more viable option.

Presently, there are some studies that assess use of novel microbial consortiums to mitigate abiotic stresses in various crops, like *Zea mays* and *Camellia sinensis* (Molina-Romero et al. 2017). Bhattacharyya et al. (2020), recently worked with plant growth-promoting properties of tea rhizobacteria and induced defense mechanism of tea. However, there needs to be an elaborative study on these prospects and the impact of these consortia on crops.

Though the field of bioremediation seems to be progressing slowly, except for the use of extremophiles, cell bioremediation is a very sustainable way to remediate the pollutants released from industrial effluents to render them less toxic. Because these extremophiles and polyextremophiles can grow in the altered microenvironment of the industrial effluents, they tend to show extraordinary efficiency in wastewater treatment. The textile industry's colored effluents need decolorization because they affect light penetration, increase salt concentrations, and change the pH of the water, making bacterial growth difficult. However, extremophiles are tolerant to such stresses and can grow under such conditions. Halotolerant and halophilic bacteria can decolorize these dyes by breaking down the azo dyes into their components (biphenyl, naphthalene, etc.) with the help of certain enzymes, such as azoreductase, laccase, etc., which render the dyes colorless. Metal biosorption, bioaccumulation, or chemical transformations of metals like copper, cobalt, nickel, and chromium are also caused by extremophilic bacteria, which are psychrotolerant, halophilic, or halotolerant, and can survive at high metal concentrations, implying that they are metallotolerant and can be used in treating metal-contaminated industrial wastewater. The various applications of these techniques have been briefly illustrated in Figure 3.4.

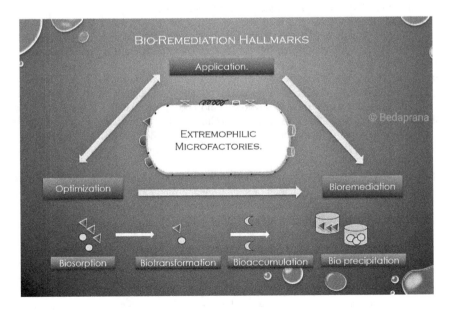

**FIGURE 3.4** Key applications of bioremediation.

The field discussed here offers great potential for biotechnological advances and sustainability. More microorganism species that have bioremediation capabilities need to be explored, and more multidisciplinary studies need to be encouraged. Progress in these two areas will bring to light the specific mechanisms and practical application approaches of using microorganisms as bioremediation to combat the serious problem of the textile industry polluting potable water with heavy metals and dyes.

## REFERENCES

Adegoke, K. A., and O. S. Bello. 2015. "Dye sequestration using agricultural wastes as adsorbents." *Water Resources and Industry* 12: 8–24. http://www.academia.edu/29211637/Dye_sequestration_using_agricultural_wastes_as_adsorbents.

Ajao, A. T., G. B. Adebayo, and S. E. Yakubu. 2011. "Bioremediation of Textile Industrial Effluent using mixed culture of *Pseudomonas aeruginosa* and *Bacillus subtilis* immobilized on agar in a Bioreactor." *Journal of Microbiology and Biotechnology Research* 1: 50–56.

Allam, Nanis G. 2017. "Bioremediation efficiency of heavy metals and azo dyes by individual or consortium bacterial species either as free or immobilized cells: a comparative study, Egypt." *Journal of Botany* 57 (3): 555–564. doi: 10.21608/ejbo.2017.689.1040.

Ali, L. 2010. "Biodegradation of synthetic dyes-a review." *Water, Air, & Soil Pollution* 213: 251–273.

Anton, A., C. Große, J. Reißmann, T. Pribyl, and D. H. Nies. 1999. "CzcD is a heavy metal ion transporter involved in regulation of heavy metal resistance in *Ralstonia* sp. strain CH34." *Journal of Bacteriology* 181 (22): 6876–6881.

Asad, S., M. A. Amoozegar, A. A. Pourbabaee, M. N. Sarbolouki, and S. M. M. Dastgheib. 2007. "Decolorization of textile azo dyes by newly isolated halophilic and halotolerant bacteria." *Bioresource Technology* 98 (11): 2082–2088.

Asghar, Anam, Abdul Aziz, Abdul Raman Wan, and Mohd Ashri Wan Daud. 2015. "Advanced oxidation processes for in-situ production of hydrogen peroxide/hydroxyl radical for textile wastewater treatment: a review", *Journal of Cleaner Production* 87 (15): 826–838.

Bano, A., J. Hussain, A. Akbar, K. Mehmood, M. Anwar, M. S. Hasni, S. Ullah, S. Sajid, and I. Ali. 2018. "Biosorption of heavy metals by obligate halophilic fungi." *Chemosphere* 199: 218–222. https://doi.org/10.1016/j.chemosphere.2018.02.043.

Bhattacharya, Amrik, Nidhi Goyal, and Anshu Gupta. 2017. "Degradation of azo dye methyl red by alkaliphilic, halotolerant Nesterenkonia lacusekhoensis EMLA3: application in alkaline and salt-rich dyeing effluent treatment." Extremophiles, *Springer*. Japan. doi: 10.1007/s00792-017-0918-2.

Bhattacharyya C, Banerjee S, Acharya U, Mitra A, Mallick I, Haldar A, Haldar S, Ghosh A, and Ghosh A. 2020. Evaluation of plant growth promotion properties and induction of antioxidative defense mechanism by tea rhizobacteria of Darjeeling, India, Scientific Reports volume 10, Article number: 15536 (2020)

Brunetti, G., K. Farrag, P. Soler-Rovira, M. Ferrara, F. Nigro, and N. Senesi. 2012. "The effect of compost and Bacillus licheniformis on the phytoextraction of Cr, Cu, Pb and Zn by three Brassicaceae species from contaminated soils in the Apulia region, Southern Italy." *Geoderma* 170: 322–330.

Chanmugathas, P., and J. M. Bollag. 1988. "A column study of the biological mobilization and speciation of cadmium in soil." *Archives of Environmental Contamination and Toxicology* 17: 229–237.

Chatha, S. A. S., M. Asgher, and H. M. N. Iqbal. 2017. "Enzyme-based solutions for textile processing and dye contaminant biodegradation—a review." *Environmental Science And Pollution* 24: 14005–14018. https://doi.org/10.1007/s11356-017-8998-1.

Chellaiah, R. 2018. "Cadmium (heavy metal) bioremediation by *Pseudomonas aeruginosa:* a minireview." *Applied Water Science* 8:154. https://doi.org/10.1007/s13201-018-0796-5.

Chen, K. C., J. Wu, D. J. Liou, and S. C. J. Hwang. 2003. "Decolorization of the textile azo dyes by newly isolated bacterial strains." *Journal of Biotechnology* 101: 57–68.

Choudhary, Madhu, Raman Kumar, Ashim Datta, Vibha Nehra, and Neelam Garg. 2017. "Bioremediation of heavy metals by microbes, bioremediation of salt affected soils: an Indian perspective." doi: 10.1007/978-3-319-48257-6_12.

Das, S., H. R. Dash, and J. Chakraborty. 2016. "Genetic basis and importance of metal resistant genes in bacteria for bioremediation of contaminated environments with toxic metal pollutants." *Applied Microbiology and Biotechnology* 100: 2967–2984. doi: 10.1007/s00253-016-7364-4.

Dekker, L., F. Arsène-Ploetze, and J. M. Santini. 2016. "Comparative proteomics of Acidithiobacillus ferrooxidans grown in the presence and absence of uranium." *Research in Microbiology* 167: 234–239. doi: 10.1016/j.resmic.2016.01.007.

De Oliveira, T. B., C. Gostinčar, N. Gunde-Cimerman, and A. Rodrigues. 2018. "Genome mining for peptidases in heat-tolerant and mesophilic fungi and putative adaptations for thermostability." *BMC Genomics* 19: 1–10. https://doi.org/10.1186/s12864-0184549-5.

De Serrano, L. O., A. K. Camper, and A. M. Richards. 2016. "An overview of siderophores for iron acquisition in microorganisms living in the extreme." *Biometals* 29: 551–571. doi: 10.1007/s10534-016-9949-x.

Dermont, G., M. Bergeron, G. Mercier, and M. Richerlaflèche. 2008. "Soil washing for metal removal: A review of physical/chemical technologies and field applications." *Journal of Hazardous Materials* 152: 1–31.

Douay, F., A. Pelfrêne, J. Planque, H. Fourrier, A. Richard, H. Roussel, and B. Girondelot. 2013. "Assessment of potential health risk for inhabitants living near a former lead smelter. Part 1: metal concentrations in soils, agricultural crops, and homegrown vegetables." *Environmental Monitoring and Assessment* 185: 3665–3680. https://doi.org/10.1007/s10661-012-2818-3.

Ehrlich, H. L. 1997. "Microbes and metals." *Applied Microbiology and Biotechnology* 48: 687–692.

El-Wakeel, Shaimaa T., R. M. Moghazy, A. Labena, and Sh. Husien. 2019. "Algal biosorbent as a basic tool for heavy metals removal; the first step for further applications." *Journal of Materials and Environmental Sciences* 10 (1): 75–87.

Equitymaster 2020, https://www.equitymaster.com/research-it/sector-info/textiles/Textiles-Sector-Analysis-Report.asp

Franke, S., G. Grass, C. Rensing, and D. H. Nies. 2003. "Molecular analysis of the copper transporting efflux system CusCFBA of *Escherichia coli*." *Journal of Bacteriology* 185(13): 3804–3812.

Gazso, G. L. 2001. "The key microbial processes in the removal of toxic metals and radionuclides from the environment." *Central European Journal of Occupational and Environmental Medicine* 7: 178–185.

Georgiou, D. G., A. Aivazidis, J. Hatiras, and K. Gimouhopoulos. 2003. "Treatment of cotton textile wastewater using lime and ferrous sulfate." *Water Research* 37 (9):2248–2250. doi: 10.1016/S0043-1354(02)00481-5.

Giovanella, Patricia, Gabriela A.L. Vieira, Igor V. Ramos Otero, Elisa Pais Pellizzer, Bruno de Jesus Fontes, and Lara D. Sette. 2019. "Metal and organic pollutants bioremediation by extremophile microorganisms." *Journal of Hazardous Materials* 382: 121024.

Gomare, S. S., and S. P. Govindwar. 2009. "*Brevibacillus laterosporus* MTCC 2298: a potential azo dye degrader." *Journal of Applied Microbiology* 106: 993–1004.

Goyal, N., S. C. Jain, and U. C. Banerjee. 2003. "Comparative studies on the microbial adsorption of heavy metals." *Advances in Environmental Research* 7: 311–319.

Grass, G., B. Fan, B. P. Rosen, K. Lemeke, H. G. Schlegal, and C. Rensing. 2001. "NreB from *Achromobacter xylosoxidans* 31A is a nickel-induced transporter conferring nickel resistance." *Journal of Bacteriology* 183 (9): 2803–2807.

He, L. M., and B. M. Tebo. 1998. "Surface charge properties of and Cu(II) adsorption by spores of the marine *Bacillus* sp. Strain SG-1." *Applied and Environmental Microbiology* 64: 1123–1129.

Hill, W. E., S. Perkins, and G. S. Sandlin. 1993. "Removal and speciation of transition metal ions textile dyeing wastewaters." *Textile Chemists and Colorists* 25: 26–27.

Huang, C., and C. P. Huang. 1996. "Application of *Aspergillus oryzae* and *Rhizopus oryzae* for Cu (II) removal." *Water Research* 30: 1985–1990.

Huang, WW., J. M. Martin, P. Seyler, J. Zhang, and X. M. Zhong. 1985. "Distribution and behaviour of arsenic in the Huang He (yellow river) estuary and Bohai sea." *Marine Chemistry* 25(1):75–79. https://doi.org/10.1016/0304-4203(88)90016-3.

Hussein, H., R. Krull, S. I. Abou El-Ela, and D. C. Hempel. 2001. "Interaction of the different heavy metal ions with immobilized bacterial culture degrading xenobiotic wastewater compounds." Presented at the IWA Conference, Berlin, Germany. 15–19.

Hussain, J., I. Hussain, and M. Arif. 2004. "Characterization of textile wastewater." *Journal of Industrial Pollution Control* 20(1): 137–144.

Hussain, J., F. U. Khan, R. Ullah, Z. Muhammad, N. U. Rehman, Z. K. Shinwari, I. U. Khan, M. Zohaib, Imad-ud-din, and A. M. Hussain. 2011. Nutrient evaluation and elemental analysis of four selected medicinal plants of Khyber Pakhtoon Khwa, Pakistan. *Pakistan Journal of Botany* 43 (1): 427–434.

Hynninen, A. 2010. "Zinc, cadmium and lead resistance mechanism in bacteria and their contribution in biosensing." ISBN 978-952-10-6263-6 (PDF).

Hynninen, A., K. Tõnismann, and M. Virta. 2010. "Improving the sensitivity of bacterial bioreporters for heavy metals." *Bioengineered Bugs* 1 (2): 132–138.

Ibrahim, A. S. S., M. A. El-Tayeb, Y. B. Elbadawi, and A. A. Al-Salamah. 2011. "Isolation and characterization of novel potent Cr(VI) reducing alkaliphilic *Amphibacillus* sp. KSUCR3 from hypersaline soda lakes." *Electronic Journal Biotechnology* 14 (4): 4. https://doi. org/10.2225/vol14-issue4-fulltext-4.

Jadhav S. U., S. D. Kalme, and S. P. Govindwar. 2008. "Biodegradation of methyl red by *Galactomyces geotrichum* MTCC 1360." *International Biodeterioration and Biodegradation* 62: 135–142.

Khehra, M. S., H. S. Saini, D. K. Sharma, B. S. Chadha, and S. S. Chimni. 2005. "Decolorization of various azo dyes by bacterial consortium." *Dyes and Pigment* 67: 5–61.

Kirk and Othmer, 1993, "Chlorous Acid, Chlorites, and Chlorine Dioxide", *Kirk Othmer Encyclopedia of Technology*, 4th edition, vol 5, pp. 968–997. John Wiley & Sons.

Kumar, Ashutosh, Anwar Alama, Deeksha Tripathi, Mamta Rani, Hafeeza Khatoon, Saurabh Pandey, Nasreen Z. Ehtesham, and Seyed E. Hasnain. 2018. "Protein adaptations in extremophiles: An insight into extremophilic connection of mycobacterial proteome." *Seminars in Cell & Developmental Biology* 84: 147–157. https://doi. org/10.1016/j.semcdb.2018.01.003.

Lade, H., S. Govindwar, and D. Paul. 2015. "Mineralization and detoxification of the carcinogenic azo dye Congo red and real textile effluent by a polyurethane foam immobilized microbial consortium in an upflow column bioreactor." *International Journal of Environmental Research and Public Health* 12: 6894–6918. https://doi.org/10.3390/ijerph120606894.

Mabrouk, M. E. M., M. A. Arayes, and S. A. Sabry. 2014. "Hexavalent chromium reduction by chromate-resistant haloalkaliphilic *Halomonas* sp. M-Cr newly isolated from tannery effluent." *Biotechnology & Biotechnology Equipment* 28 (4): 659–667. https://doi.org/10.1080/ 13102818.2014.937092.

Mahmood, R., F. Sharif, S. Ali, and M. U. Hayyat. 2013. "Bioremediation of textile effluent by indigenous bacterial consortia and its effects on *Zea mays* L. CV C1415." *Journal of Animal & Plant Sciences* 23 (4): 1193–1199.

Molina-Romero, Dalia, Antonino Baez, Verónica Quintero-Hernández, Miguel Castañeda-Lucio, Luis Ernesto Fuentes-Ramírez, María del Rocío Bustillos-Cristales, Osvaldo Rodríguez-Andrade, Yolanda Elizabeth Morales-García, Antonio Munive, and Jesús Muñoz-Rojas. 2017. "Compatible bacterial mixture, tolerant to desiccation, improves maize plant growth." https://doi.org/10.1371/ journal.pone.0187913.

Marchler-Bauer, A., J. B. Anderson, M. K. Derbyshire, C. DeWeese-Scott, N. R. Gonzales, M. Gwadz, L. Hao, S. He, D. I. Hurwitz, J. D. Jackson, Z. Ke, D. Krylov, C. J. Lanczycki, C. A. Liebert, C. Liu, F. Lu, S. Lu, G. H. Marchler, M. Mullokandov, J. S. Song, N. Thanki, R. A. Yamashita, J. J. Yin, D. Zhang, and S. H. Bryant. 2007. "CDD: A conserved domain database for interactive domain family analysis." *Nucleic Acids Research* 35 (suppl 1): D237–D240.

Markandeya, T., S. P. Shukla, and D. Mohan. 2007. "Toxicity of disperse dyes and its removal from wastewater using various adsorbents: a review." *Research Journal of Environmental Toxicology* 11: 72–89. https://scialert.net/fulltext/?doi=rjet.2017.72.89.

Marques, C. R. 2018. "Extremophilic microfactories: applications in metal and radionuclide bioremediation." *Frontiers in Microbiology* 9: 1191. doi: 10.3389/ fmicb.2018.01191.

Marques, C. R., A. L. Caetano, A. Haller, F. Goncalves, R. Pereira, and J. Rombke. 2014. "Toxicity screening of soils from different mine areas — a contribution to track the sensitivity and variability of *Arthrobacter globiformis* assay." *Journal of Hazardous Material* 274: 331–341. doi: 10.1016/j.jhazmat.2014.03.066.

Moosvi, S., H. Keharia, and D. Madamwar. 2005. "Decolourization of textile dye Reactive Violet 5 by a newly isolated bacterial consortium RVM 11.1." *World Journal of Microbiology and Biotechnolgy* 21: 667–672. https://doi.org/10.1007/ s11274-004-3612-3

Navarro, C. A., D. von Bernath, and C. A. Jerez. 2013. "Heavy metal resistance strategies of acidophilic bacteria and their acquisition: importance for biomining and bioremediation." *Biological Research* 46: 363–371. doi: 10.4067/S0716-97602013000400008.

Nies, D. H. 2003. "Efflux-mediated heavy metal resistance in prokaryotes." *FEMS Microbiology Reviews* 27 (2-3): 313–339.

Peeples, Tonya L. 2014. "Bioremediation using extremophiles." *Microbial Biodegradation and Bioremediation* 251–268. http://dx.doi.org/10.1016/B978-0-12-800021-2.00010-8.

Prasad, M. N. V., K. S. Sajwan, and R. Naidu. 2006. "Trace elements in the Environment: Biogeochemistry, Biotechnology and Bioremediation." *Experimental Agriculture*. CRC Press, Boca Raton, FL. Pp. 726. ISBN 1-566-70-685-8.

Rahman, M. S., and K. V. Sathasivam. 2015. "Heavy metal adsorption onto *Kappaphycus* sp. from aqueous solutions: the use of error functions for validation of isotherm and kinetics models." *BioMed Research International* 126298.

Sarkar, S., A. Banerjee, U. Halder, R. Biswas, and R. Bandopadhyay. 2017. "Degradation of synthetic azo dyes of textile industry: a sustainable approach using microbial enzymes." *Water Conservation Science and Engineering* 2 (4): 121–131. https://doi.org/10.1007/ s41101017-0031-5.

Scott, J. A., and A. M. Karanjkar. 1992. "Repeated cadmium biosorption by regenerated *Enterobacter aerogenes* biofilm attached to activated carbon." *Biotechnology Letters* 14: 737–740.

Senan, R., and T. Abraham. 2004. "Bioremidiation of Textile azo dyes by aerobic bacterial consortium aerobic degradation of selected azo-dye by bacterial consortium" *Biodegradation* 15 (4):275–80. doi: 10.1023/B:BIOD.0000043000.18427.0a.

Sheth, N. T., and S. R. Dave. 2009. "Optimisation for enhanced decolourization and degradation of Reactive Red BS C.I. 111 by *Pseudomonas aeruginosa* NGKCTS", *Biodegradation* 20 (6): 827–36. doi: 10.1007/s10532-009-9270-2. Epub 2009 Jun 11.

Siddiquee, S., R. Kobun, and S. Saallah. 2015. "Heavy metal contaminants removal from wastewater using potential filamentous fungi biomass: a review." *Journal of Microbiology and Biochemical technology.* doi: 10.4172/1948-5948.1000243.

Singh, A., and A. K. Singh. 2017. "Haloarchaea: worth exploring for their biotechnological potential." *Biotechnology Letter* 39: 1793–1800. doi: 10.1007/s10529-017-2434-y.

Srivastava, P., J. Bragança, S. R. Ramanan, and M. Kowshik. 2013. "Synthesis of silver nanoparticles using haloarchaeal isolate *Halococcus salifodinae* BK3." *Extremophiles* 17: 821–831. doi: 10.1007/s00792-013-0563-3.

Sumathi, S., G. I. R. Shruthy, S. Vinothini, M. Jayaprakashvel, and G. Kalavathy. 2018. "Bioremediation of texile dye (Magenta MP) using halotolerant marine bacteria." *IJEDR* 6 (2). ISSN: 2321–9939.

Sungur, Sana, and Faith Gülmez. 2015. "Determination of metal contents of various fibers." *Journal of Spectroscopy* 2015 (3). http://dx.doi.org/10.1155/2015/640271.

Tomova, I., M. Stoilova-Disheva, and E. Vasileva-Tonkova. 2014. "Characterization of heavy metals resistant heterotrophic bacteria from soils in the Windmill Islands region, Wilkes Land, East Antarctica." *Polish Polar Research* 35 (4): 593–607. https://doi. org/10. 2478/popore-2014-0028.

Verma, Pradeep, and Datta Madamwar. 2003. "Decolourization of synthetic dyes by a newly isolated strain of Serratia marcescens." *World Journal of Microbiology and Biotechnology* 19: 615–618.

Volesky, B., and Z. R. Holan. 1995. "Biosorption of heavy metals." *Biotechnology Progress* 11: 235–250.

Wang, Y., J. Guo, and R. Liu. 2001. "Biosorption of heavy metals by bacteria isolated from activated sludge." *Applied Biochemistry and Biotechnology* 91–93: 171–184.

Yuyao, Jin, Yaning Luan, Yangcui Ning, and Lingyan Wang. 2018. "Effects and mechanisms of microbial remediation of heavy metals in soil: a critical review." *Applied Science* 8: 1336. doi:10.3390/app8081336.

Zhang, F., A. Yediler, X. Liang, and A. Kettrup. 2004. "Effects of dye additives on the ozonation process and oxidation by products: a comparative study using hydrolysed CI reactive red 120." *Dyes and Pigments* 60: 1–7.

Zhang, X., X. Liu, Y. Liang, F. Fan, X. Zhang, and H. Yin. 2016. "Metabolic diversity and adaptive mechanisms of iron- and/or sulfur-oxidizing autotrophic acidophiles in extremely acidic environments." *Environmental Microbiology* 8: 738–751. doi: 10.1111/1758-2229.12435.

Zhao, M., P.-F. Sun, L.-N. Du, G., Wang, X.-M. Jia, and Y.-H. Zhao. 2014. "Biodegradation of methyl red by *Bacillus* sp. strain UN2: decolorization capacity, metabolites characterization, and enzyme analysis." *Environmental Science and Pollution* 21: 6136–6145.

# 4 From the Green Revolution to the Green Chemistry Revolution

## In Pursuit of a Paradigm Shift in Agricultural Sustainability

*Usha Rao and Aiden Saul*

## CONTENTS

## 4.1 INTRODUCTION

Technological innovations in agriculture have increased crop yields, reduced world hunger, and supported the large population growth of the previous century. However, sustainable agricultural practices have become an urgent need due to exploding population growth, increased demand for food, the lack of available new land for farming, and environmental problems including deforestation, water resource depletion, pesticide overuse, topsoil loss, and air and water pollution. Humanity has crossed or is approaching several "planetary boundaries," in particular with global climate change, biodiversity loss, land use, freshwater quality and quantity, ocean acidification, and changes to the biogeochemical cycles of nitrogen and phosphorus

(Rockstrom et al. 2009). Consumption in developed countries, such as those in the Organisation for Economic Cooperation and Development (OECD), exerts a global ecological footprint that is many times that of non-member countries (Anastas and Lankey 2002). As BRICS countries (Brazil, Russia, India, China and South Africa) begin to achieve a higher standard of living, the challenge becomes one of how to reproduce the wealth of the developed world without an equally damaging ecological footprint, and in the face of climate change's accelerating and detrimental effects.

The world population now stands at 7.6 billion and is expected to reach 9.1 billion by 2050 (Alexandratos and Bruinsma 2012), requiring 25–100% more food production (Hunter et al. 2017). Approximately 70% of global freshwater goes toward supporting agriculture, with India and China alone dedicating over a trillion cubic meters of water for agriculture in 2015 (Ritchie n.d.). As the effects of climate change intensify, water availability for agriculture is a pressing concern.

Pests and pathogens consume 5–20% of major cereal crops each year, costing approximately 470 billion USD globally (Sharma et al. 2017). Climate change is estimated to increase crop losses by a further 10–25% per °C of mean global surface temperature warming, propelled by rising insect metabolic rates and insect population growth (Deutsch et al. 2018). To prevent pest-induced losses, four million metric tons of pesticides are used each year worldwide (FAOSTAT n.d.). Pesticide manufacture is a 15 billion USD industry in the United States alone, employing 12,000 workers (IBIS World n.d.). Many pesticides, herbicides, and fungicides pose a threat to ecosystems due to their propensity to bioaccumulate, bioconcentrate, and biomagnify (e.g., Deribe et al. 2013). Due to large-scale spraying, over 98% of insecticides and 95% of herbicides reach non-target species and geochemical reservoirs (Miller 2004). As insects develop pesticide resistance, the effectiveness of pesticide formulations becomes time-delimited, necessitating an arms race against insect evolution. Due to their carcinogenic, mutagenic, and teratogenic effects, pesticides affect the health of workers, especially those working without adequate protective gear, as is a common practice in many parts of the world. A reported 25 million agricultural workers are exposed to pesticide poisoning each year (Alavanja 2009). However, the effects of pesticide, herbicide, and fungicide use are not limited to agricultural workers. Approximately 50 million Americans use agriculturally polluted groundwater as their primary drinking source (Ward et al. 2000). As awareness of the hazards associated with pesticides increases, the urgent need for more benign replacements for traditional pesticides has grown.

The overuse of fertilizer has led to numerous environmental problems such as water and air pollution, eutrophication and hypoxia, climate change exacerbation, and ozone depletion (e.g., Carpenter et al. 1998; Rao et al. 1999; Galloway et al. 2008; Reay et al. 2012). Farm runoff threatens freshwater streams and lakes (Haith and Dougherty 1976). Changing land-use patterns, along with the need for roads and housing for new agricultural communities, contribute to deforestation and further intensify global change by removing sinks for carbon.

Food waste causes a significant ecological problem in developed countries. American consumers waste 422 g of food per capita per day, which amounts to

30 million acres of farmland (8% of cropland) used merely to produce food waste. This waste consumes 4.2 trillion gallons of irrigation water, 780 million pounds of pesticide, and 5.6 billion pounds of fertilizer (Conrad et al. 2018), and causes needless atmospheric and water pollution.

Sustainable agricultural use that does not deplete or damage biogeochemical resources is dependent on more benign pesticide and fertilizer design; the incorporation of smart agricultural technologies; the innovative use and recycling of water; reductions in pollution, soil losses, energy use; along with building adaptive resilience to climate change into agricultural practice.

The United Nations General Assembly adopted 17 sustainable development goals (SDGs) in 2015 (Table 4.1) for attainment by 2030. Many of the SDGs deal directly or indirectly with agriculture, mandating a new approach to efficient, non-polluting agriculture that is resilient to climate change. These practices must support viable livelihoods for farmers, eradicate hunger, deliver nutritious food, mitigate water overuse and pollution, minimize air pollution, especially from greenhouse gases, reduce energy waste and promote a transition to sustainable energy systems, and reduce deforestation and topsoil loss. These and other sustainability goals can only be met with the involvement of the chemical enterprise, in particular, by integrating the SDGs into the chemical supply chain (Noce 2018).

---

**TABLE 4.1**
**United Nations Sustainable Development Goals in the 2030 Agenda for Sustainable Development**

Goal 1.　No Poverty
Goal 2.　Zero Hunger
Goal 3.　Good Health and Well-Being
Goal 4.　Quality Education
Goal 5.　Gender Equality
Goal 6.　Clean Water and Sanitation
Goal 7.　Affordable and Clean Energy
Goal 8.　Decent Work and Economic Growth
Goal 9.　Industry, Innovation, and Infrastructure
Goal 10. Reduced Inequality
Goal 11. Sustainable Cities and Communities
Goal 12. Responsible Consumption and Production
Goal 13. Climate Action
Goal 14. Life below Water
Goal 15. Life on Land
Goal 16. Peace and Justice Strong Institutions
Goal 17. Partnerships to Achieve the Goal

Many of the Sustainable Development Goals (SDG's) developed by the General Assembly of the United Nations in 2015 pertain directly or indirectly to agriculture (https://www.un.org/development/desa/disabilities/envision2030.html).

---

## 4.2   GREEN CHEMISTRY

The chemical industry is one of the largest in the world, supporting 120 million jobs and adding over a trillion USD to the global GDP (International Council of Chemical Associations n.d.). A recent global inventory determined that over 350,000 chemicals and chemical mixtures are registered for use (Wang et al. 2020). Approximately 95% of all manufactured goods and products rely on chemistry, with applications in agriculture, pharmaceuticals, transportation, energy, and information technology. Given chemistry's role in manufacturing nearly every known convenience, it is unsurprising that the use of chemical products is the biggest source of pollution on Earth. The benefits of chemistry are well-known, but the detrimental effects of, and widespread societal exposure to, human-made chemical substances are equally important.

Until the past few decades, chemists had a limited role in providing solutions to the problems associated with the unconstrained use of chemicals, beyond quantifying the extent and magnitude of the pollution after release. But the toxicity and polluting properties of hazardous substances derive from the molecular properties of chemical substances, and so therefore must the solutions. As humanity's need to achieve sustainability gains urgency, the design framework of green chemistry, embodied in the 12 principles listed in Table 4.2, shows promise as a provider of viable alternatives. The field of green chemistry was initiated by the United States Environmental Protection Agency (U.S. EPA) and the United States National Science Foundation (NSF) to encourage researchers to develop more benign chemical products and processes. The American Chemical Society's Green Chemistry Institute brings together stakeholders to include systems thinking in chemical research and education. Several advances in green chemistry methodologies such as streamlined and efficient synthetic pathways, innovative development of enzymes and other catalysts, and the adoption of environmentally friendly solvents, have been promising

---

## TABLE 4.2
## The 12 Principles That Constitute the Framework of Green Chemistry

Principle 1.  Prevent Waste
Principle 2.  Atom Economy
Principle 3.  Less Hazardous Synthesis
Principle 4.  Design Benign Chemicals
Principle 5.  Benign Solvents and Auxiliaries
Principle 6.  Design for Energy Efficiency
Principle 7.  Use of Renewable Feedstocks
Principle 8.  Reduce Derivatives
Principle 9.  Catalysis (vs. Stoichiometric)
Principle 10. Design for Degradation
Principle 11. Real-Time Analysis for Pollution Prevention
Principle 12. Inherently Benign Chemistry for Accident Prevention

Source: www.acs.org/greenchemistry.

---

(Anastas and Beach 2007). The Green Chemistry Educators Network was established with the goal of modernizing the traditional chemistry curriculum which had not "changed much for 50 years or more" (Ritter 2016) to include earth and environmental science, toxicology, systems thinking, and cradle-to-grave lifecycle analysis to prepare students to undertake the challenge of conducting chemistry for a sustainable future. The technologies highlighted by the Presidential Green Chemistry Challenge Awards, designed to encourage the development of sustainable chemical practices, had resulted in "826 million pounds of hazardous chemicals and solvents eliminated each year—enough to fill almost 3,800 railroad tank cars or a train nearly 47 miles long; 21 billion gallons of water saved each year—the amount used by 820,000 people annually; 7.8 billion pounds of carbon dioxide equivalents released to air eliminated each year—equal to taking 810,000 automobiles off the road" by 2016 (U.S. EPA n.d.).

## 4.3 THE PATH FORWARD

The framework of green chemistry offers a potential path forward toward attaining agricultural sustainability. This study provides a geochemical perspective into promising new lines of research, examines challenges to wider adoption of green chemistry, and offers recommendations to increase the integration of green chemistry and sustainable chemistry into agricultural practices that are resilient in the face of climate change.

### 4.3.1 SEMIOCHEMICALS

The problems associated with the use of conventional organic and inorganic pesticides include point-source and nonpoint-source pollution, toxicity to humans and other organisms, lack of species selectivity, development of resistance by target species, and persistence in the environment (Ward et al. 2000; Miller 2004; Alavanja 2009; Deribe et al. 2013). The use of synthetic insect sex and aggregation pheromones to disrupt the reproductive cycles of target species by delaying or preventing mating is emerging as a substitute for traditional pesticides, for instance, in lepidopteran pest control (Gut et al. 2004). Mating disruption via pheromones has been used in the control of the codling moth (*Cydia pomonella*) in apple and pear trees, and the grapevine moth (*Lobesia botrana*) in grapevines (Witzgall et al. 2010). The pheromones are target-specific, minimizing toxic exposure to non-target species, and preventing the widespread ecological damage associated with conventional pesticides. Due to the high potency and biodegradability of biopesticides as compared to traditional pesticides, far lower concentrations are required and result in lower levels of pesticide resistance development. The use of semiochemicals also reduces toxic chemical exposure to farmers and consumers.

The major drawback to the use of human-made semiochemicals has been the high cost of synthesizing pheromones due to the required purity and stereospecificity of the products, especially in use as sex attractants in traps where isomeric impurities as low as 1% can impede use (Turczel et al. 2019). Green chemistry techniques have been used to lower the cost of synthesis and to find safer synthetic pathways such as *in vivo* synthesis of pheromone precursors from inexpensive fatty-acid feedstock (Knipple et al. 2000). The development of highly selective catalyst systems in olefin metathesis reactions offers great promise in pest control by requiring lower energy inputs, catalyst loading, and overall cost of synthesis, and by achieving close to 100% atom economy in reaction classes such as ring opening cross-metathesis and ring-opening metathesis polymerization (Turczel et al. 2019).

There is enormous untapped potential for future semiochemicals research, such as the case of the gypsy moth (*Lymantria dispar*) which is one of the 100 most destructive invasive species worldwide. Pheromone disruption would also be helpful in the suppression of the Swede midge (*Contarinia nasturtii* Kieffer) whose larvae are a threat to cruciferous plants such as cabbage, cauliflower, and broccoli. The midge is newly invasive to North America, where it has the potential to cause widespread *Brassica* oilseed and cruciferous vegetable crop losses. There are no available organic management systems to protect affected crops. Traditional pyrethroid, carbamate, and organophosphate pesticides, although toxic to the midge, do not reach midges in the field (Cornell University Swede Midge Information Center for the United States n.d.). The high cost of synthesizing its pheromone, a blend of three chiral components (2S,9S-diacetoxyundecane, 2S,10S-diacetoxyundecane, and 2S-acetoxyundecane in a 1:2:0.02 ratio) has been a roadblock to using mating disruption. In the case of other pests such as the invasive stink bug *Bagrada hilaris* that destroys cruciferous crops, recent studies have used 2-D NMR spectroscopy to identify the attractant brassicadiene released by the plant as the specific diterpene responsible for drawing pest insects to the crops (Arriola et al. 2020). Although the complex structure of brassicadiene may prove a barrier for synthesizing large volumes of the compound cost-effectively, the identification of the relevant diterpene offers a new avenue for research - breeding cruciferous vegetables without brassicadiene.

### 4.3.2 Biopesticides

Natural products derived from plants, bacteria, and derived proteins offer a vast array of potential resources for synthetic chemists beyond the usual approaches such as screening compound libraries and synthesizing newer analogs of existing compounds. Nature produces approximately 170 billion tons of biomass annually, of which humans use only 3.5% for our needs. The use of biomass-derived feedstocks would protect biodiversity by providing an impetus to preserve natural species, exhibit reduced environmental persistence, and protect human and ecosystem health. Combining the use of different natural pesticides, especially in mixtures not commonly occurring in nature or in a certain region, may prove particularly effective and is an approach that warrants future study.

The neem tree (*Azadirachta indica*), native to India and used for insecticidal purposes in granaries by Indian farmers for thousands of years, is the classic example of a successful biopesticide that has led to hundreds of formulations. The neem tree was labeled "the single best source of biopesticides around" by the U.S. National Research Council panel (Stone 1992). Neem seeds, leaves, and bark contain several limonoids, in particular the tetranortriterpenoid azadirachtin which acts on over 200 insect species. Azadirachtin blocks the hormone ecdysone that regulates molting and prevents maturation of immature larval insects. It is also a feeding deterrent, growth inhibitor, mating disruptor, and/or sterility inducer in different populations (Stone 1992). Azadirachtin degrades rapidly under environmental conditions, is non-carcinogenic, and poses low or no toxicity to birds and mammals (Mordue and Blackwell, 1993). While azadirachtin was finally synthesized after 22 years of experimentation, the process involved 64 chemical transformations (Veitch et al. 2007), which suggests that the compound will continue to be extracted from the neem tree for the foreseeable future.

Fermentation of soil microorganisms is a time-tested method of deriving useful substances, dating back to the discovery of penicillin. Corteva Agriscience developed Spinosad from the microorganism *Saccharopolyspora spinosa*, extracted from a Caribbean soil sample found to be active on mosquito larvae. The spinosyns, a large family of unique compounds that are produced from *Saccharopolyspora* fermentation, contain a polyketide-derived tetracyclic core attached to two saccharides. Spinosad, the combination of two principal fermentation factors, produces neurotoxicity in insects resulting in paralysis and death. Spinosad is believed to inhibit nicotinic and γ-aminobutyric acid (GABA) receptor function through a novel neurotoxicity mechanism (Kirst 2010). It targets a range of insects with an efficiency comparable to that of pyrethroids, and is also effective against external parasites of livestock, pets, and humans (head lice). Unlike traditional pesticides, Spinosad, now sold commercially in over 80 preparations and a cornerstone of integrated pest management (IPM) techniques, offers many environmental advantages over traditional pesticides. It absorbs onto soil and only minimally leaches into groundwater; shows low volatilization; photochemically degrades and does not persist in the environment; and is generally more target-selective, leaving over 70% of beneficial insects and other non-target insect species unharmed. Spinosad is also non-carcinogenic and of low toxicity in mammals, and a safer alternative for farmers. These characteristics led to Spinosad being recognized by the U.S. EPA's Presidential Green Chemistry Challenge Award (U.S. EPA n.d.). Spinetoram, an analogue of Spinosad, also won a Green Chemistry Challenge Award in 2008.

In a departure from being soil-microbe derived like many biopesticides, Vestaron Corporation's Spear®-T insecticide draws its inspiration from the venomous funnel-web spider, *Hadronyche versuta*. Spear-T is based on the action of the GS-omega/kappa-Hxtx-Hv1a peptide which blocks calcium and potassium ion channels in the nervous systems of a wide range of arthropod pest insects. Vestaron Corporation genetically engineered yeast to produce the peptide in large quantities. Due to its biodegradability and no appreciable toxicity to mammals, fish, or bees, Spear-T received a Presidential Green Chemistry Award (Bomgardner 2017).

The case of the spotted lantern fly in Eastern North America offers an emerging case study on the use of biopesticides. The spotted lantern fly, *Lycorma delicatula*, an invasive species from China was first identified in Pennsylvania in 2014 and has since been spreading to neighboring states (New York State Integrated Pest Management Program n.d.). It is a polyphagous planthopper that preferentially feeds on *Ailanthus altissima* (tree of heaven), but also on 70 other common tree species from 25 families. Pennsylvania's grape, apple, and other tree fruit, hardwood logging, and nursery industries that annually contribute USD 18 billion to the state's

economy are at significant risk from this species. The introduction of natural predators is being researched as a potential solution, but introduced predator species often
pose ecosystem disruptive challenges of their own. A recent finding offers a glimmer
of promise in the treatment of this species that has no known insect or avian predators in Eastern North America. Clifton et al. (2019) reported that two fungal entomo-
pathogens native to the region, *Batkoa major* and *Beauveria bassiana* were causing a
coepizootic, successfully attacking an outbreak population of the spotted lantern fly.
The U.S. EPA already approves a *B. bassiana* spray for the treatment of bedbug colo-
nies, such as in the formulation Aprehend®. Research should be conducted to assess
whether a combination spray made of the two fungal pathogens could treat spotted
lantern fly infestation, if the two fungi are found to be environmentally benign. The
spray could be aerially applied on vulnerable cropland, perhaps with the additional
use of drones, robots, and on-ground spraying for targeted applications that exploit
the potential of smart technology.

### 4.3.3 GENETICALLY MODIFIED ORGANISMS

Since the introduction of genetically modified (GM) tobacco in 1986 and Roundup®
(glyphosate) Ready soybeans in 1996, GM crops have become commonly used in
agriculture. A meta-analysis of 147 studies found that GM crops had "reduced pesticide use by 37%, increased crop yields by 22%, and increased farmer profits by 68%"
(Klumper and Qaim 2014). In a special issue on GM crops in the journal *Nature*,
Gilbert (2013) reported that GM technology increased agricultural productivity by
98 billion USD while preventing the spraying of 473 million kilograms of traditional
pesticides. It must be noted, however, that public skepticism and fear about GM crops
has not abated in the decades of use.

Nearly all current GM crops are grown to be paired with a glyphosate-based
herbicide. Glyphosate is a non-selective, systemic, widely used, post-emergence
herbicide that acts by inhibiting the chloroplast enzyme, 5-enolpyruvylshikimate-
3-phosphate synthase (EPSPS), which is needed to synthesize the amino acids
phenylalanine, tyrosine, and tryptophan via the shikimate pathway in plants and
some bacteria (Heap and Duke 2018). The relatively cheap price of glyphosate led
to it being used, sometimes indiscriminately, in soy, corn, cotton, canola, sugar
beet, and other crops worldwide. It has sometimes replaced traditional practices
of ploughing and tilling, crop rotation, and the use of a broad range of herbicides to discourage weed growth. Due to this intensive use, 38 species of agricultural weeds in 37 countries have developed resistance to glyphosate (Heap and
Duke 2018), with nearly half of the resistant species found in the United States.
Growers have traditionally moved to new herbicides to overcome the development
of evolutionary resistance, progressing from triazine and dinitroaniline, ALS and
ACCase inhibitors, protoporphyrinogen oxidase (PPO) inhibitors, to glyphosate.
Heap and Duke (2018) report that herbicide discovery efforts have stalled over
several decades, and more concerningly, there are "no new herbicide modes of
action and few new chemistries," which makes the emergence of glyphosate-
resistant weeds a challenging problem. Total herbicide use related to GM crops
is still on the rise in the United States and predicted to reach 3.5 kg/hectare in

2025 (Mortensen et al. 2012). However, it is likely that we are reaching the useful lifespan of the current generation of herbicides in GM and non-GM crops, after decades of use and evolutionary resistance. To continue to be a viable weed control method, glyphosate and other herbicides must become part of herbicide combinations used in rotation, as part of a well-designed IPM scheme instead of remaining stand-alone strategies.

### 4.3.4 FERTILIZER ADDITIVES

Reducing and optimizing the use of fertilizers in agriculture is a pressing concern, as the global consumption of nitrogen, phosphorus (expressed as phosphate, $P_2O_5$), and potassium (expressed as potassium oxide, $K_2O$) reached nearly 187 Tg in 2017 (FAO n.d.). For 2020, the UN Food and Agriculture Organization (FAO) anticipated 202 Tg of use, with approximately half of the total fertilizer consumption coming from nitrogen fertilizer.

Sustainable nitrogen fertilizer use is an urgent problem – anthropogenic perturbations to the nitrogen biogeochemical cycle exceed even those to the global carbon cycle (Davidson et al. 2011). The Haber-Bosch synthesis of ammonia is one of the most important scientific innovations of the past century, because it lowered energy needs and thus the cost of fertilizer. Eighty percent of the total nitrogen manufactured by the Haber-Bosch process goes toward fertilizer production (Galloway et al. 2008). Smil (1999) estimated that fertilizer-grown food supported 40% of the world's population at the end of the twentieth century. In fact, without the use of synthetic nitrogen fertilizer, the global population would have peaked at 3.84 billion (Our World in Data n.d.). While this smaller population size would have exerted a smaller footprint on the planet, as things stand, nearly one in two humans owes their existence to the Haber-Bosch synthesis.

Fertilizer nitrogen, usually applied as urea or ammonia, offers clear economic advantages to food production, but its use carries many environmental drawbacks: water pollution and eutrophication, tropospheric and stratospheric pollution, climate change, and stratospheric ozone depletion (Carpenter et al. 1998; Rao et al. 1999; Galloway et al. 2008; Reay et al. 2012). Nitrogen fertilizer production is estimated to increase two- to three-fold by mid-century (Erisman et al. 2008), further exacerbating the problem. Unfortunately, nutrient use efficiency (NUE), the fraction of all nitrogen inputs (mainly via fertilizer but also manure, biological fixation, and deposition) to the nitrogen in food (harvested crops, dairy and meat products) is extremely low, averaging 20%, and showing few improvements over the years (Smil, 1999; Zhang et al. 2015).

Plant growth is severely limited by nitrogen deficiencies due to nitrogen's vital role in the synthesizing chlorophyll, protein, and DNA, but maintaining a stable concentration of bioavailable soil nitrogen is challenging due to the many microbial processes that bring about soil nitrogen transformation. Overuse of nitrogen fertilizer, a strategy to compensate for losses, leads to the aforementioned pollution. For example, there has been a noted increase of fertilizer runoff to the Mississippi-Atchafalaya watershed, the third largest drainage basin in the world, over the past few decades. Nitrate and phosphate runoff from midwestern US farms enters the

Gulf of Mexico via this drainage basin, leading to algal blooms, eutrophication, biodiversity loss, the formation of the largest hypoxic zone in the western Atlantic Ocean, anoxia, and the seasonal creation of a dead zone in the Gulf of Mexico (Ribuado et al. 2001). In fact, two in three coastal zones in the United States are moderately to severely impacted by nutrient inputs, with 300 hypoxic (<2 mg/L dissolved oxygen) zones existing along the US coastline, with a third of all US streams and two-fifths of lakes experiencing ecological impairment from nutrient inputs (Davidson et al. 2011).

Additionally, up to 60% of non-$CO_2$ anthropogenic greenhouse gas emissions, especially of nitrous oxide, $N_2O$, are attributed to the agricultural sector, which is the fourth largest global warming contributor (Smith 2017). Nitrous oxide has a greenhouse warming potential that is 310 times that of carbon dioxide on a per-molecule basis, in addition to being a stratospheric ozone destroyer. Pollution from NOx causes biodiversity losses, with the selective loss of nitrogen sensitive native species over losses of invasive species (Davidson et al. 2011). The World Health Organization estimates that one in eight premature human deaths worldwide can be attributed to NOx pollution, from respiratory problems, cardiovascular disease, SIDS, cancer, and birth defects.

Chemical agents that inhibit soil nitrogen losses can reduce the severity of environmental effects. Nitrification inhibitors and urease inhibitors play an important role in retaining fertilizer in the soil root zone, promoting nitrogen retention and plant growth, and reducing nitrate leaching and the release of NOx emissions from agricultural soil. Wolt (2004) reported in a meta-analysis of agricultural soils in the midwestern United States that the application of nitrapyrin (2-chloro-6-(trichloromethyl)pyridine) which inhibits the urease enzyme in *Nitrosomonas* bacteria, increased mean corn crop yield by 7% and nitrogen retention by 28%. Nitrate leaching was reduced by 16% and greenhouse gas emissions were reduced by 51% on average. The recent development of a technique to micro-encapsulate nitrapyrin in a low-volatility form (Wilson et al. 2013) led to the creation of the Instinct® nitrification inhibitor by Dow Agrosciences. Instinct was proven to improve the efficiency of soil nitrogen uptake and to reduce NOx emissions (Burzaco et al. 2014) and received a Presidential Green Chemistry Award. The U.S. EPA reported that this nitrification inhibitor produced 50 million additional bushels of corn and reduced $CO_2$ equivalent emissions by 664,000 metric tons over five years (EPAb n.d.). In 2018, Koch Agronomics introduced CENTURO®, formulated with a new nitrification inhibitor, pronitridine, a complex reaction product consisting of oligomeric reacted species of N-cyanoguanidine. A recent study found that pronitridine increased yield by 7% over nitrapyrin when applied in the Fall, and gave similar yields when applied immediately prior to Spring planting (Singh and Nelson 2019). It is essential to select the appropriate nitrogen fertilizer and additives for a given set of environmental conditions such as climate, soil classification, nutrient availability, timing of fertilizer additives, and water management techniques. This will become particularly important as the full effects of climate change unfold – for instance, the U.S. Department of Agriculture reported that climate change-related rainfall and flooding reduced the harvests of certain crops by up to 20% in the United States in 2019 (USDA NASS n.d.). On a global scale, the market growth potential for new and effective formulations of nitrogen fertilizer loss inhibitors in agriculture is immense.

### 4.3.4.1 Phosphorus Fertilizers

Phosphorus is an essential nutrient for all life forms due to its myriad roles in cellular protein formation, nucleic acid synthesis, metabolic energy transfer, root growth and stability, photosynthesis, seed formation, disease resistance, among other functions (Cockefair, 1931). Although soil phosphorus concentrations can reach 1000 mg/kg, less than 2.5% of the phosphorus is bioavailable to plant roots, with the vast majority being bound up in metal-cation complexes that are refractory in soil. Due to the poor lability and mobility in soil, phosphorus fertilizer is needed to support plant growth. The intensive use of phosphorus in agriculture means that fertilizer phosphorus now dominates the biogeochemical cycle for the element (Oelkers and Valsami-Jones 2008).

Due to the depletion of phosphate rock sources, phosphorus is a non-renewable resource, and is expected to reach "peak phosphorus" levels by 2033, and be depleted within 100 years (Cordell et al. 2011). In addition to phosphorus fertilizer pollution such as eutrophication and the creation of hypoxic and anoxic zones in water bodies, the impending shortage of phosphorus provides an additional impetus for finding novel ways to maximize the use of phosphorus fertilizer.

Phosphorus use efficiency in agriculture can be improved by targeted use of the biological potential within the rhizosphere and the natural beneficial relationships between plants and other organisms. Mycorrhizae, rhizobacteria, and nitrogen-fixing rhizobia can all support plant growth and nutrient uptake and repel pathogens (Etesami and Maheshwari, 2018). Soil bacterial species that grow in the rhizosphere and enhance plant growth and vitality are collectively known as plant growth-promoting rhizobacteria (PGPR). These include *Pseudomonas putida*, *Azospirillum fluorescens*, and *Azospirillum lipoferum*. Common nitrogen-fixing bacteria in legumes such as *Allorhizobium*, *Azorhizobium*, *Bradyrhizobium*, and *Rhizobium* are also included in the PGPR group. In a natural setting, PGPR perform many useful functions such as enhancing nutrient availability, increasing root growth and morphology, and creating symbiotic relationships with plants (Vessey 2003). Mycorrhizae, a mutual symbiotic association between a fungus and a plant in the rhizosphere, are associated with increased nutrient uptake and promotion of plant growth and vitality. The fungi colonize the plant's roots, intracellularly with the hyphae of endomycorrhizal fungi penetrating the plant cell wall as in arbuscular mycorrhizal fungi, or extracellularly as ectomycorrhizal fungal species. The arbuscular mycorrhizal fungi transfer nutrients such as phosphorus, zinc, and copper via mycorrhizal mycelial networks to plants, receiving carbon in exchange; stabilize soil aggregates; and alleviate biotic and abiotic stresses on vegetation (Smith and Read 2014).

The simultaneous application of phosphorus-solubilizing microorganisms and arbuscular mycorrhizal fungi was shown to have synergistic effects in plant phosphorus uptake (Babana and Antoun 2006). In some field trials, crop yield increased by up to 20% (Jones and Oburger 2011). But in other field studies, PGPR were not found to be consistently effective, perhaps related to the efficiency of colonizing plant roots (Benizri et al. 2001), a process that is controlled by the expression of a range of bacterial traits and specific genes (Lugtenberg et al. 2001). In studies of arbuscular mycorrhizal associations with plant roots, even when phosphorus uptake by roots was enhanced, plant growth could yet be suppressed (Smith and Read 2014), possibly due to variability in the expression of symbiosis-associated genes depending on fungal species, plant genotype, and environmental conditions, indicating that further investigation is warranted. The use of phosphate-solubilizing microorganisms such as arbuscular mycorrhizal fungi and PGPR may be useful in increasing phosphorus use efficiency due to their role in producing inorganic and organic acids, increasing surface area and uptake by roots, improving stress tolerance, and building symbiotic relationships with host plants (Etesami and Maheshwari 2018), but further studies on the biochemical mechanisms involved are needed to optimize their use in agriculture.

### 4.3.5 NANOMATERIALS

Agricultural nanotechnology represents a novel approach to increasing crop yield and promoting plant stress resistance and growth. While the nano-agriculture field is still in its infancy, nanomaterials have been explored in studies of seed germination, plant growth, pathogen diagnosis, environmental condition detection, and improving plant stress responses to biotic and abiotic factors, among others (Zhao et al. 2020). Nanocarriers can derive from natural sources such as carbohydrates (e.g., chitosan, gum arabic, cellulose, starch, guar gum), proteins (e.g., zein, lectin), and lipids (phospholipids, cholesterol, triglycerides, lecithins) (Oliveira et al. 2018). Nanostructured formulations provide controlled and targeted release of ingredients to plants. Due to their small size, large surface area, and available binding sites, nanoparticles are good carriers for bioactive materials, e.g., plasmid DNA and double-stranded RNA (Kwak et al. 2019) or can be encapsulated with pesticides, nutrients, and uptake enhancers (Zhao et al. 2020).

Metal nanopesticides have been investigated for potential use in improving plant response to biotic stresses such as pathogen and pest invasion. Ocsoy et al. (2013) demonstrated that silver nanocrystals grown on graphene oxide composites significantly decreased the activity of the pathogenic bacterium *Xanthomonas perforans*, which reduces tomato yields by up to 50%. Silver nanoparticles applied at a concentration of 150 mg/mL reduced plant parasitic root-knot nematode burdens in bermudagrass by 92% within four days (Cromwell et al. 2014).

An EPA-approved herbicide, the naturally occurring tobacco mild green mosaic virus, was also found to be an effective carrier to deliver nematicides. This plant virus self-assembles into a $300 \times 18$ nm nanorod with a hollow inner channel, and presents solvent-exposed carboxylate groups from Glu, Tyr, and Asp side chains, which form good targets for bioconjugation or metal ion binding. The plant viral nanomaterial showed good soil mobility and efficacy, making it a suitable carrier for antihelminitic drugs such as crystal violet to treat nematode infections (Chariou and Steinmetz 2017).

Zein nanoparticles, developed from corn protein, can be effective carriers for agrochemicals, lowering production costs due to their biological sourcing, and potentially reducing environmental contamination. Zein nanoparticles encapsulating geraniol (3,7-dimethylocta-trans-2,6-dien-1-ol) and R-citronellal (3,7-dimethyloct-6-en-1-al), components of the essential oil of citronella, were effective in repelling the *Tetranychus urticae* Koch mite, a polyphagous and destructive agricultural pest (Oliveira et al. 2018). Unlike botanical insect repellents which often undergo degradation by sunlight, heat, and microbes, zein nanoparticles loaded with the repellents can offer stability and increased delivery of product to target sites.

Nanotechnology can also deliver traditional fertilizers, nutrient uptake enhancers, and enzymes to plants. Avellan et al. (2017) demonstrated that gold nanoparticles applied to leaves reached the mesophyll, roots, and even the rhizospheric soil. Kottegoda et al. (2017) have developed a nitrogen nanofertilizer using urea-hydroxyapatite nanoparticles with a high surface area and an aspect ratio around 10. Foliar application of zinc nanoparticles was found to increase the uptake of existing soil phosphorus by 11% in legume and cereal crops, without the need for applied fertilizer (Raliya et al. 2018). The authors reported that zinc nanoparticles provide this required micronutrient and mobilize soil phosphorus by increasing the activities of phosphatase, phytase, and other phosphorus-solubilizing enzymes that require zinc as a cofactor. Zinc nanoparticles also have the ability to enhance crop nutritional value, for instance, lycopene content in tomato plants (Raliya et al. 2018). Enhanced phosphorus uptake was also noted with the use of magnetite, $Fe_3O_4$, nanoparticles, with the hypothesis that nanoparticles contribute to enhanced root exudation and soil acidification, leading to increased bioavailability (Zahra et al. 2015).

Nanomaterials may also be useful in increasing plant resistance to abiotic stresses, such as those from extreme temperatures, drought, salinity, heavy metal toxicity, and oxidative stress, all of which adversely affect plant growth and development (Atkinson and Urwin 2012). Nanoparticles of cerium(IV) oxide, fullerenes such as C60, and magnetite enhance plant ability to scavenge reactive oxygen species; some nanoparticles trigger upregulation of genes that enhance antioxidant and stress tolerance related metabolic pathways (Zhao et al. 2020).

Ecosystem studies of biomagnification, biotoxicity, and other parameters related to the use of nanomaterials in agriculture are needed to understand and constrain the pathways, cycling, sinks, and ecological hazards of introduced nanomaterials in natural settings. While nanomaterials offer great promise in effective delivery of required substances to crops, their application is hindered by the absence of better methodologies for detecting these nanomaterials and their byproducts in ecosystems. We also lack a detailed risk assessment framework for the environmental impact and human health consequences arising from the use of agricultural nanomaterials. A recent critical assessment of nanopesticides and nanofertilizers against their conventional analogues found them to increase efficacy by 30%, but also reported that comprehensive studies of the interaction of nanomaterials with the environment are sorely lacking (Kah et al. 2018). A green life-cycle approach for nanochemicals that integrates botanical formulations with the use of biodegradable, non-toxic matrices (Oliveira et al. 2018), lower use of solvents and lower energy consumption in production, and benign, rapidly degraded end-products should be the development goal for this new field.

## 4.4 IMPLEMENTING GREEN CHEMISTRY IN AGRICULTURE – BARRIERS AND SOLUTIONS

Agriculture is the largest industry on Earth, employing one in three humans world-wide. Although consumers tend to picture farms as existing in a pastoral, age-old setting, agriculture is becoming an intensively managed system worldwide, with a heavy reliance on synthetic fertilizer, crop protection products, and GM technology. Food needs are increasing globally, driven by a growing population and rising per capita consumption and meat consumption. The ability to produce more food per unit area of land is threatened by changing climate, with agriculture itself being a major contributor to GHG emissions, which need to be reduced by as much as 80% to avert the worst outcomes of climate change (Foley et al. 2011). Agricultural production must adapt to new challenges such as water shortages and pesticide resistance. Agriculture's environmental footprint has become unsustainable, as we recognize that the Green Revolution and its promise of cheap food has come at a growing cost to human and environmental health, and the abandonment of traditional, ecologically friendlier practices adapted to regional environmental conditions in favor of unsustainable monocultures. Increases in food production must be coupled with significant decreases in environmental impact, necessitating nothing short of a paradigm shift in how humans farm.

As the examples in this chapter demonstrate, the green chemistry framework can bring about that much-needed paradigm shift by advocating for a more benign, responsible, and sustainable approach to creating, using, and recycling agrochemicals. In fact, although green chemistry's contributions to the U.S. pharmaceutical industry are rightly lauded, the greatest potential for the use of a green chemistry framework lies in the agricultural realm, due to its significantly larger global scale of operations, diversity of products, and the number of people employed and affected. Currently, the primary adopters of green chemistry are in the United States, United Kingdom, Japan, Italy, and Australia, raising the question of how to create widespread adoption within these and other countries. Some financial disincentives, oversight constraints, and educational shortcomings which can act as barriers to green chemistry adoption are discussed here.

Aggregate data from five leading agribusinesses with active discovery programs revealed that the overall cost of agrochemicals, from first synthesis to development, registration, and commercialization, increased steadily from 1995 to reach 286 million USD in the 2010–2014 period, the latest period surveyed. The development period likewise increased by over 35%, with increases attributed in part to regulatory requirements in the form of field studies and toxicity and environmental safety data collection (Phillips McDougall AgriService n.d.). Chemical enterprises interested in embracing green chemistry need extra assurance of increased profitability that offsets the cost of retooling existing methodology and re-building physical spaces such as factories. Product price is critical in farming, unlike areas such as pharmaceuticals where consumers are willing to pay a higher price for products. Most farmers work with a small profit margin and fixed costs related to land rent, labor, and equipment. This dichotomy between the cost involved in developing new, greener agrochemicals and farmer preference for low costs and familiar products presents a challenge. For instance, biopesticides only made up 8% of pesticide sales in North America in 2012, with the bioinsecticide market in particular stagnating in recent years (Bomgardner 2017). Surprisingly, this stagnation followed on the heels of significant investment in biopesticides by major agribusinesses, such as the purchase of AgraQuest by Bayer, perhaps because early green formulations were less effective and more expensive, leading to farmer skepticism and unwillingness to experiment with new pest control methods. Involving stakeholders from the earliest stages may help the industry better engage farmers in green products and convince them of efficacy. Governments and NGO's should help farmers transition to newer and safer chemicals use by providing subsidies and tax incentives, grants, and ongoing training.

Agrochemicals are highly regulated, and rightly so, given their toxicity to farm workers and consumers and their propensity to cause environmental damage. However, the length and expense of meeting regulatory barriers in a small market with low profit margins adversely affects the biologically derived products market. Hunter et al (2017) point out that agricultural activities are exempted from many environmental regulations, and propose that governmental subsidies to agriculture be closely tied to proof of meeting environmental standards. Priority setting by governmental agencies should identify, quantify, and link food production goals and environmental goals. Increased governmental support of green chemistry practices in the form of tax incentives and regulations will help to grow the field in the face of other challenges.

Agricultural sustainability is an inter-disciplinary, multi-stakeholder endeavor, requiring better integration and communication between producers, consumers, and policymakers to meet both environmental and economic needs. An electronic, searchable database of green chemistry applications and methods, integrating approaches from industry, government, and academia, available freely on a global scale, and pro-actively disseminated to stakeholders is a primary need in this area. For instance, the UN Environment Programme's Life Cycle Initiative, a tool to assess the ecotoxicity of chemical substances, is a partnership between governmental bodies, nongovernmental agencies, and academia. Some resources are available independently, such as the database of pheromone and semiochemicals on the website pherobase.com that lists 3,500 chemical substances with mass spectra and synthesis pathways. The NGO Clean Production Action provides tools such as the GreenScreen benchmarking tool and strategies to assist a chemical enterprise's transition to safer chemicals. But, these remain fragmented and isolated efforts with little to no coordination at national or international levels. Since food is widely exported between nations, a trans-national, trans-boundary effort is what is needed. Technology improvements in data acquisition, storage, and communication over the past decade now make better data sharing and cooperative efforts on a large scale both possible and relatively inexpensive. Geographic Information Systems (GIS) also make pollutant tracking and pattern establishment feasible. An overarching database that combines all green chemistry agricultural methodologies, built from a multi-stakeholder, global perspective by an international agency such as the UN FAO would be a significant asset in promoting agricultural sustainability choices.

Green chemistry is by nature interdisciplinary. Within the chemistry educator community, there is rising awareness that systems thinking must be integrated into the chemistry curriculum from the beginning (Pernaa and Aksela 2011; Blatti et al. 2019). In the American Chemical Society's flagship environmental journal, *Environmental Science and Technology*, Matus et al. (2012) identify the lack of "the ability to think on a more global systems level" as a significant barrier to the adoption of green chemistry in the chemical enterprise, and state that chemistry curricula "have remained remarkably constant since the end of World War." Chemists are not typically exposed to fields such as toxicology, earth and environmental science, and ecology in their education, and students in these disciplines likewise have only cursory exposure to the valuable tools offered by synthetic and analytical chemistry, although cutting-edge research often takes place at the interface between disciplines. Young scientists will need a broader technical background to enter the marketplace ready to solve environmental problems, or at least to not add more toxic products to the inventory. On the research front, journals such as the American Chemical Society's *Environmental Science and Technology* and the Royal Society of Chemistry's *Journal of Green Chemistry* are expanding the reach of green chemistry to all scientists.

We must rely on green chemistry principles in the creation of new and environmentally protective agricultural products and practices, recognize farmlands as complex ecosystems, pursue circular economies and closed-loop systems, and create life cycle assessments for a comprehensive cradle-to-grave analysis of the production, use, and disposal of agrochemicals to promote better environmental decision making. Integrating knowledge from diverse disciplines such as chemistry, data management, ecology, earth and environmental science, technology, toxicology, and inviting the full participation of the farming community, industry, academia, governmental and non-governmental bodies, will be required to turn into reality the dream of agriculture that can feed billions while reducing the ecological footprint on a burdened planet at a climate-tipping point.

## REFERENCES

Alavanja, M.C.R. 2009. Introduction: pesticides use and exposure extensive worldwide. *Reviews on Environmental Health* 24(4): 303–309.

Alexandratos, N. and J. Bruinsma. 2012. World agriculture towards 2030/2050: the 2012 revision. ESA Working paper No. 12-03. Rome, FAO.

Anastas, P.T. and E.S. Beach. 2007. Green chemistry: the emergence of a transformative framework. *Green Chemistry Letters and Reviews* 1(1): 9–24.

Anastas, P.T. and R. Lankey. 2002. Sustainability through green chemistry and engineering. *Advancing Sustainability through Green Chemistry and Engineering*, ed. Lankey, R. L. and P.T. Anastas, 1–11. ACS Publications 823. American Chemical Society: Washington, DC.

Atkinson, N. and J. Urwin. 2012. The interaction of plant biotic and abiotic stresses: from genes to the field. *Journal of Experimental Botany* 63(10): 3523–3543.

Arriola, K., S. Guarino, C. Schlawis, M.A. Arif, S. Colazza, E. Peri, et al. 2020. Identification of brassicadiene, a diterpene hydrocarbon attractive to the invasive stink bug *Bagrada hilaris*, from volatiles of cauliflower seedlings, *Brassica oleracea* var. botrytis. *Organic Letters* 22(8): 2972–2975.

Avellan A., F. Schwab, A. Masion, P. Chaurand, D. Borschneck, V. Vidal, et al. 2017. Nanoparticle uptake in plants: gold nanomaterial localized in roots of *Arabidopsis thaliana* by X-ray computed nanotomography and hyperspectral imaging. *Environmental Science and Technology* 51: 8682–8691.

Babana, A.H. and H. Antoun. 2006. Effect of Tilemsi phosphate rock-solubilizing microorganisms on phosphorus uptake and yield of field-grown wheat (*Triticum aestivum L.*) in Mali. *Plant Soil* 287: 51–58.

Benizri, E., E. Baudoin, A. Guckert. 2001. Root colonization by inoculated plant growth promoting rhizobacteria. *Biocontrol Science and Technology* 11(5): 557–574.

Blatti, J.L., J. Garcia, D. Cave, F. Monge, A. Cuccinello, J. Portillo, et al. 2019. Systems thinking in science education and outreach toward a sustainable future. *Journal of Chemical Education* 96(12): 2852–2862.

Bomgardner, M.M. 2017. Spider venom: an insecticide whose time has come? *Chemical and Engineering News* 95(11): 30–31.

Burzaco, J.P., I.A. Ciampitti, T.J. Vyn. 2014. Nitrapyrin impacts on maize yield and nitrogen use efficiency with spring-applied nitrogen: Field studies vs. meta-analysis comparison. *Agronomy Journal* 106: 753–760.

Carpenter, S.R., N.F. Caraco, D.L. Correll, R.W. Howarth, A.N. Sharpley, V.H. Smith. 1998. Nonpoint pollution of surface waters with phosphorus and nitrogen. *Ecological Applications* (8): 559–568.

Chariou, P.L. and N.F. Steinmetz. 2017. Delivery of pesticides to plant parasitic nematodes using tobacco mild green mosaic virus as a nanocarrier." *ACS Nano* 11(5): 4719–4730.

Clifton, E.H., L.A. Castrillo, A. Gryganskyi, A.E. Hajek. 2019. A pair of native fungal pathogens drives decline of a new invasive herbivore. *Proceedings of the National Acadamy of Sciences of the United States of America* 116(19): 9178–9180.

Cockefair, E.A. 1931. The Role of Phosphorus in the Metabolism of Plants. *American Journal of Botany* 18: 582.

Conrad, Z., M.T. Niles, D.A. Neher, E.D. Roy, N.E. Tichenor, L. Jahns. 2018. Relationship between food waste, diet quality, and environmental sustainability. *PLoS One* 13(4): e0195405.

Cordell, D., A. Rosemarin, J.J. Schroder, A.L. Smit. 2011. Towards global phosphorus security: a systems framework for phosphorus recovery and reuse options. *Chemosphere* 84(6): 747–758.

Cornell University Swede Midge Information Center for the United States. n.d. http://web.entomology.cornell.edu/shelton/swede-midge/management.html.

Cromwell, W.A., J. Yang, J.L. Starr, Y.K. Jo. 2014. Nematocidal effects of silver nanoparticles on root-knot nematode in bermudagrass. *Journal of Nematology* 46(3): 261–266.

Davidson, E.A., M.B. David, J.N. Galloway, C.L. Goodale, R. Haeuber, J.A. Harrison, et al. 2011. Excess nitrogen use in the U.S. environment: trends, risks, and solutions. *Issues in Ecology.* 15: 1–6.

Deribe, E., B.O. Rosseland, R. Borgstrom, B. Salbu, Z. Gebremariam, E. Dadebo, et al. 2013. Biomagnification of DDT and its metabolites in four fish research species of a tropical lake. *Ecotoxicology and Environmental Safety.* 95: 10–18.

Deutsch, C.A., M. Tigchelaar, D.S. Battisti, S.C. Merrill, R.B. Huey, R.L. Naylor. 2018. Increase in crop losses to insect pests in a warming climate. *Science.* 361(6405): 916–919.

Erisman, J., M. Sutton, J. Galloway, Z. Klimont, W. Winiwarter. 2008. How a century of ammonia synthesis changed the world. *Nature GeoScience* 1: 636–639.

Etesami, H. and D.M. Maheshwari. 2018. Use of plant growth promoting rhizobacteria (PGPRs) with multiple plant growth promoting traits in stress agriculture: Action mechanisms and future prospects. *Ecotoxicology and Environmental Safety* 156(30): 225–246.

FAO. n.d. Food and Agriculture Organization of the United Nations. World fertilizer trends and outlook to 2020 summary report. http://www.fao.org/3/a-i6895e.pdf

FAOSTAT. n.d. Food and Agriculture Organization of the United Nations. Pesticides use. http://www.fao.org/faostat/en/#data/RP/visualize

Foley, J.A., N. Ramankutty, K.A. Brauman, E.S. Cassidy, J.S. Gerber, M. Johnstone, et al. 2011. Solutions for a cultivated planet. *Nature* 478: 337–342.

Galloway, J.N., A.R. Townsend, J.W. Erisman, M. Bekunda, Z. Cai, J.R. Freney, et al. 2008. Transformation of the nitrogen cycle: recent trends, questions, and potential solutions. *Science* 320: 889–892.

Gilbert, N. 2013. Case studies: a hard look at GM crops. *Nature* 497: 24–26.

Gut, L.J., L.L. Stelinski, D.R. Thompson, J.R. Miller. 2004. Behaviour-modifying chemicals: prospects and constraints in IPM. In: *Integrated Pest Management: Potential, Constraints, and Challenges*. eds. Koul O., Dhaliwal G.S., Cuperus G. 73–121. Wallingford, UK. CABI Publishing.

Haith, A.D. and J.V. Dougherty. 1976. Nonpoint source pollution from agricultural runoff. *Journal of the Environmental Engineering Division*. 102(5): 1055–1069.

Heap, I. and S.O. Duke. 2018. Overview of glyphosate-resistant weeds worldwide. *Pest Management Science* 74(5): 1040–1049.

Hunter, M.C., R.G. Smith, M.E. Schipanski, L.W. Atwood, D.A. Mortensen. 2017. Agriculture in 2050: recalibrating targets for sustainable intensification. *BioScience* 67(4): 386–391.

IBIS World, n.d. Pesticide manufacturing industry in the US – Market research report. https://www.ibisworld.com/united-states/market-research-reports/pesticide-manufacturing-industry/

International Council of Chemical Associations. n.d. The Global Chemical Industry: catalyzing growth and addressing our world's sustainability challenges. https://icca-chem.org/wp-content/uploads/2020/09/Catalyzing-Growth-and-Addressing-Our-Worlds-Sustainability-Challenges.pdf

Jones, D.L. and E. Oburger. 2011. Solubilization of phosphorus by soil microorganism. *Soil Biology* 26(1): 169–198.

Kah, M., R.S. Kookana, A. Gogos, T.D. Bucheli. 2018. A critical evaluation of nanopesticides and nanofertilizers against their conventional analogues. *Nature Nanotechnology* 13: 677–684.

Kirst, H.A. 2010. The spinosyn family of insecticides: realizing the potential of natural products research. *Journal of Antibiotics* 63(1): 101–111.

Klumper, W. and M. Qaim. 2014. A meta-analysis of the impacts of genetically modified crops. *PLoS One* 9(11): e111629.

Knipple, D.C., W. L. Roelofs, C.-L. Rosenfield, P. Marsella-Herrick, S.J. Miller. 2000. In vivo synthesis in yeast of insect sex pheromone precursors: an alternative synthetic route for the production of environmentally benign insect control agents. In: *Green Chemical Syntheses and Processes*, eds. Anastas, P.T., Heine, L.G., and Williamson, T.C. 33–42, *ACS Symposium Series* 767, American Chemical Society.

Kottegoda, N., C. Sandaruwan, G. Priyadarshana, S. Siriwardhana, U.A. Rathnayake, D.M. Berugoda Arachchige, et al. 2017. Urea-hydroxyapatite nanohybrids for slow release of nitrogen. *ACS Nano* 11(2): 1214–1221.

Kwak, S.Y., T.T.S. Lew, C.J. Sweeney, V.C. Koman, M.H. Wong, K. Bohmert-Tatarev, et al. 2019. Chloroplast-selective gene delivery and expression in planta using chitosan-complexed single-walled carbon nanotube carriers. *Natural Nanotechnology* 14(5): 447–455.

Lugtenberg, B.J.J., L. Dekkers, G.V. Bloemberg. 2001. Molecular determinants of rhizosphere colonization by *Pseudomonas. Annual Review of Phytopathology* 4(4): 343–350.

Oliveira, J.L., E.V.R. Campos, L. Fernandes. 2018. Recent developments and challenges for nanoscale formulation of botanical pesticides for use in sustainable agriculture. *Journal of Agricultral and Food Chemistry* 66(34): 8898–8913.

Matus K.J., W.C. Clark, P.T. Anastas, J.B. Zimmerman. 2012. Barriers to the implementation of green chemistry in the United States. *Environmental Science and Technology* 46(20): 10892–9.

Miller, G.T. 2004. *Sustaining the Earth: An Integrated Approach.* 211–216. Brooks Cole.

Mordue A. J. and A. Blackwell. 1993. Azadirachtin: an update. *Journal of Insect Physiology* 39: 903–924.

Mortensen, D.A., J.F. Egan, B.D. Maxwell, M.R. Ryan, R.G. Smith. 2012. Navigating a critical juncture for sustainable weed management. *BioScience* 62(1): 75–84.

New York State Integrated Pest Management Program. n.d. Spotted lanternfly. https://nysipm.cornell.edu/environment/invasive-species-exotic-pests/spotted-lanternfly/

Noce, A.M. 2018. How chemistry can help meet the UN's sustainable development goals. *Chemical and Engineering News* 96(22): 43.

Ocsoy, I., M.L. Paret, M.A. Ocsoy, S. Kunwar, T. Chen, M. You, et al. 2013. Nanotechnology in plant disease management: DNA-directed silver nanoparticles on graphene oxide as an antibacterial against *Xanthomonas perforans. ACS Nano* 7(10): 8972–8980.

Oelkers, E. H. and E. Valsami-Jones. 2008. Phosphate mineral reactivity and global sustainability. *Elements* 4: 83–87.

Our World in Data. n.d. https://ourworldindata.org/grapher/world-population-with-and-without-fertilizer

Pernaa, J. and M. Aksela. 2011. Learning organic chemistry through a study of semiochemicals. *Journal of Chemical Education* 88(12): 1644–1647.

Phillips McDougall AgriService, n.d. https://croplife.org/wp-content/uploads/2016/04/Cost-of-CP-report-FINAL.pdf

Raliya, R., V. Saharan, C. Dimkpa, P. Biswas. 2018. Nanofertilizer for precision and sustainable agriculture: current state and future perspectives. *Journal of Agricultural and Food Chemistry* 66(26): 6487–6503.

Rao, U., D.J. Hollander, T. Sirivedhin, K.A. Gray. 1999. Comparative isotopic, elemental, and molecular analyses of the seasonal changes in DOM and POM in an anthropogenically influenced riparian wetland. *Proceedings of the American Society of Limnology and Oceanography ASLO Bulletin* 8(2–3).

Reay, D.S., E.A. Davidson, K.A. Smith, P. Smith, J.M. Melillo, F. Dentener, et al. 2012. Global agriculture and nitrous oxide emissions. *Nature Climate Change* 2: 410–416.

Ribuado, M.O., R. Heimlich, R. Claassen, M. Peters. 2001. Least-cost management of non-point source pollution: source reduction versus interception strategies for controlling nitrogen loss in the Mississippi Basin. *Ecological Economics* 37(2): 183–197.

Ritchie, H. n.d. Water use and stress. https://ourworldindata.org/water-use-stress

Ritter, S.K. 2016. Green chemistry celebrates 25 years of progress. *Chemical and Engineering News* 94(27): 22–25.

Rockstrom, J., W. Steffen, K. Noone, A. Persson, C.F. Stuart, E.F. Lambin, et al. 2009. A safe operating space for humanity. *Nature* 461(7263): 472–500.

Sharma, S., R. Kooner, R. Arora. 2017. Insect pests and crop losses. In: *Breeding Insect Resistant Crops for Sustainable Agriculture.* Eds. Arora, R. and S. Sandhu, 45–66. Springer Nature, Singapore.

Singh, G. and K. Nelson. 2019. Pronitridine and nitrapyrin with anhydrous ammonia for corn. *Journal of Agricultural Science* 11(4): 13–24.

Smil, V. 1999. Nitrogen in crop production: an account of global flows. *Global Biogeochemical Cycles* 13(2): 647–662.

Smith, K.A. 2017. Changing views of nitrous oxide emissions from agricultural soil: key controlling processes and assessment at different spatial scales *Europ. Journal of Soil Science* 68(2): 137–155.

Smith, S. and D. Read. 2014. Mycorrhizal symbiosis (3). *Elsevier and Academic.*

Stone, R. 1992. A biopesticidal tree begins to blossom. *Science* 255(5048): 1070–1071.

Turczel, G., E. Kovacs, G. Merza, P. Coish, P.T. Anastas, R. Tuba. 2019. Synthesis of semi-ochemicals via olefin metathesis. *ACS Sustainable Chemistry and Engineering* 7(1): 33–48.

USDA NASS. n.d. United States Department of Agriculture National Agricultural Statistics Service Crop Production 2019 Summary. https://www.nass.usda.gov/Publications/Todays_Reports/reports/cropan20.pdf

U.S. EPA. n.d.a Information about the green chemistry challenge. https://www.epa.gov/greenchemistry/information-about-green-chemistry-challenge

U.S. EPA, n.d.b Green Chemistry Challenge Winners. https://www.epa.gov/greenchemistry/green-chemistry-challenge-winners#main-content

Vessey, J.K. 2003. Plant growth promoting rhizobacteria as biofertilizers. *Plant and Soil* (255): 571–586.

Veitch, G.E., E. Beckmann, B.J. Burke, A. Boyer, S.L. Maslen, S.V. Ley. 2007. Synthesis of azadirachtin: a long but successful journey. *Angewandte Chemie International Edition in English* 46(40): 7629–7632.

Wang, Z., G.W. Walker, D.C.G. Muir, K. Nagatani-Yoshida. 2020. Toward a global understanding of chemical pollution: a first comprehensive analysis of national and regional chemical inventories. *Environmental Science and Technology* 54(5): 2575–2584.

Ward, M.G., S.J. Weigel, S.K. Maxwell, K.P. Cantor, R.S. Miller. 2000. Identifying populations potentially exposed to agricultural pesticides using remote sensing and a geographic information system. *Environmental Health Perspectives.* 108(1): 5–12.

Wilson, S.L., R.E. Boucher, S.M. Ferguson. 2013. Enhanced nitrification inhibitor composition. *U.S. Patent 8377.* 849B.

Witzgall, P., P. Kirsch, A. Cork. 2010. Sex pheromones and their impact on pest management. *Journal of Chemical Ecology* (36): 80–100.

Wolt, J.D. 2004. A meta-evaluation of nitrapyrin agronomic and environmental effectiveness with emphasis on corn production in the Midwestern USA. *Nutrient Cycling in Agroecosystems* 69: 23–41.

Zahra, Z., M. Arshad, R. Rafique, A. Mahmood, A. Habib, I. A. Qazi, et al. 2015. Metallic nanoparticle ($TiO_2$ and $Fe_3O_4$) application modifies rhizosphere phosphorus availability and uptake by *lactuca sativa*. *Journal of Agricultural and Food Chemistry 63*: 6876– 6882.

Zhang, X., E.A. Davidson, D.L. Mauzerall, T.D. Searchinger, P. Dumas, and Y. Shen. 2015. Managing nitrogen for sustainable development. *Nature.* 528: 51–59.

Zhao, L., L. Lu, A. Wang, H. Zhang, M. Huang, H. Wu, et al. 2020. Nano-biotechnology in agriculture: use of nanomaterials to promote plant growth and stress tolerance. *Journal of Agricultural and Food Chemistry* 68(7): 1935–1947.

# Section II

## Green Business, Banking, and Technology

# 5 The Power of Markets

*Reduce Climate Risks While Creating Inclusive Prosperity*

*Janée N. Burkhalter and Carolin D. Schellhorn*

## CONTENTS

## 5.1 INTRODUCTION: BACKGROUND AND DRIVING FORCES

Global warming (NASA, n.d.) is posing growing threats to individuals and communities that require timely solutions (Leiserowitz et al. 2019). Global trends of rising inequality are similarly well established and reinforced by the fact that the poorest and most marginalized segments of society are the ones most exposed to accelerating climate change and climate injustice (Islam and Winkel 2017; Lewis 2014; OECD 2015). In order to avoid the worst consequences from accelerating climate change for social stability and public health (CDC 2019), solutions must be found that substantially reduce climate risks while simultaneously addressing economic and social inequities thereby creating inclusive prosperity. To successfully reduce climate risks, we need climate-aware, resourceful, and healthy communities. However, public health depends critically on the mitigation of local and global climate risks (Podesta 2019). It is because of this mutual dependence that these dual goals must be addressed simultaneously and expeditiously in a broad-based, large-scale effort. This chapter outlines important conditions that would help our market-based economy achieve these goals within a broader reorientation of economic thinking.

An obvious first step is to re-envision a healthy economy. Globally, the goal has long been economic growth in the context of financial stability (Dudley 2011). With this mindset, Gross Domestic Product (GDP), a measure of a country's economic output, is the standard marker of success. The fair allocation of the rewards for producing the output, however, has never been a top leadership priority, which explains the presence of marginalized communities that suffer from restricted access to society's vital systems. In a re-imagined healthy economy, egregious inequity and marketplace vulnerability would be abolished (Demangeot et al. 2013). Individuals would not encounter situations in which they are at risk for harm or unfair treatment in health, housing, education and employment. Instead of focusing on the growth of economic output, an economy's success may instead be measured by the country's ability to meet basic human needs (Greve 2017; Maslow 1943) while adhering to planetary boundaries (Raworth 2017; Rockström et al. 2009). Thus, measures of society's well-being, such as the Social Progress Index (SPI), may provide more relevant markers of success. SPI captures the extent to which a country addresses the social and environmental needs of its residents and includes "sub-indicators concerning elements such as water, political rights, education and health" (Greve 2017, p. 1011). Like GDP, metrics similar to those provided by the SPI will have to be tracked regularly and publicly.

As climate change accelerates and the economic transition out of carbon-intensive assets and production processes into low-carbon and net-zero industries gathers speed, new disruptions and inequities are likely to compound already existing conditions of social and racial injustice. Universal access to healthcare, safe housing, clean energy, healthy food, relevant education, and other vital social systems is far from guaranteed. The need to address these problems in a timely fashion is thus becoming ever more urgent. The required solutions will have to be broad-based and characterized by multi-stakeholder engagement aimed at reducing the risks faced by individuals who experience marketplace vulnerability (Demangeot et al. 2013).

Markets are able to deliver the required large-scale solutions as long as they operate within an equitable legal and regulatory framework (ELRF) that is supportive. This chapter will outline several desirable characteristics of an ELRF. We then discuss three powerful market-based approaches that fit well within this framework and can help markets achieve the dual goals of reducing climate risks while creating inclusive prosperity: (1) carbon pricing coupled with a socially aware distribution of the proceeds; (2) sustainable resource allocations that recognize opportunity costs of financial capital, social and cultural capital as well as planetary capital and (3) partnerships between the public sector and the private sector to support transformative research, development and low-carbon innovation within the context of social and racial equity.

Thus, the traditional global goals of financial stability and economic growth must be embedded in a broader set of objectives that recognize the need to simultaneously reduce climate risks and stabilize communities in the increasingly difficult effort to secure human existence in the context of public and planetary health. Accelerating climate change coupled with rising threats of pandemics pose a monumental challenge that requires a concerted large-scale effort by humanity to reduce greenhouse gas emissions to net zero while protecting human communities, ecosystems and wildlife. Marginalized communities and people uprooted by migration, whose access to society's vital systems is restricted, are more likely to suffer more from infectious disease and violence, thus increasing risks to public health and economic stability. Consumption patterns, production and distribution processes will have to adjust now to avoid much more abrupt and disruptive adjustments later on. Vulnerable communities with meager prospects for equitable access to resources are neither able nor willing to support this massive effort. Yet, everyone's participation in mitigating and adapting to growing climate and related risks is essential, given limited time and constraints on available resources. Therefore, the widespread building of healthy and resilient communities must be prioritized in the fight against climate change, just as timely and successful climate action is a prerequisite for safeguarding public health and community resilience. Markets can drive progress toward achieving these dual goals as long as they operate within an ELRF that is enforced.

## 5.2   TOWARD AN ELRF

Markets must operate based on a set of essential principles and require certain conditions in order to function in a way that protects individuals from marketplace vulnerability (Demangeot et al. 2013; Korten 2010). Creating these conditions is impossible without an ELRF and investment in the maintenance of a competent public sector capable of enforcing equitable laws and regulations. To build an inclusive, prosperous society while simultaneously reducing climate risks requires healthy, resilient and culturally diverse communities, workplaces and markets. Strong labor and anti-discrimination laws and environmental laws are absolute necessities. A recent initiative by the Labor and Worklife Program of Harvard Law School, *Clean Slate for Worker Power*, outlines new labor law proposals that promise to strengthen the labor movement around racial and social equity thus building a fairer economy and inclusive prosperity (Bahn 2020). However, ELRF and law enforcement are unlikely to be sufficient given the enormity of the pre-existing systemic problems associated with social and climate injustice (Daly et al. 2017; Krakoff 2019; McIntosh et al. 2020). Lasting progress will require improvements in the functioning of our social, financial and political systems.

Civil rights organizations, for instance, are important complements to effective legal institutions and can help disseminate relevant information that is required in an ELRF. Recently, the United States National Association for the Advancement of Colored People (NAACP) had to warn local chapters about efforts by the fossil fuel industry to promote its interests at the expense of renewable energy (Patterson 2019; Penn 2020). While local chapters gain donations and in return advocate on behalf of donors' interests, "this advocacy has come at the expense of the Black neighborhoods, which are more likely to have polluting power plants and are less able to adapt to climate change," (Penn 2020). Information and education are prerequisites for resilience, and also for society's ability to drive timely and necessary systemic change across all economic sectors.

For the efficient and effective allocation of resources, markets rely on the public disclosure and dissemination of material financial information. In that regard, the United States Securities and Exchange Commission (SEC) has an important role as part of an ELRF. The US SEC has a responsibility to regulate the public reporting of climate-related financial risks. These include physical risks to production and distribution facilities, regulatory and reputation risks as well as risks to public health from harm as a result of climate injustice. The SEC's existing requirements allow companies to provide information in line with the recommendations of the Task Force on Climate-related Financial Disclosures (TCFD), but the Commission could do more to enforce existing requirements and ask for more detailed reporting (GAO 2018; SEC 2010).

As monetary authority and guardian of the stability of the financial system, the United States Federal Reserve must provide a reliable economic anchor for the ELRF. Thus the "Fed," too, must consider social justice and climate risks in its monetary policy and supervisory activities. It is becoming clear that ignoring these issues could jeopardize not only the United States central bank's independence, but also financial system stability and, with that, the stability of our society (Honohan 2019; Schellhorn 2020). For these reasons, and to support relevant emerging asset markets, a strong case can be made for including certified green bonds and other socially responsible bonds in the asset portfolios of central banks (Fender et al. 2019). Beyond researching and addressing the complex and dynamic problem of climate change with all of its implications for climate justice and social change within the United States Federal Reserve System, Bolton et al. (2020) argue that the Federal Reserve must assume responsibilities for policy coordination domestically as well as internationally. The massive scale of the challenge will require unprecedented collaboration and partnerships between government entities and a wide range of organizations in the private sector to establish and support the necessary ELRF and its enforcement.

## 5.3 CARBON PRICING SCHEMES WITH PROCEED DISTRIBUTION

While it is clear that, beyond an ELRF, markets need the pricing of carbon dioxide emissions to charge emitters for the human, social and physical costs to society from climate change and its consequences, there is no agreement on which method of carbon pricing is best (Kramer 2019). Several jurisdictions have imposed, or have begun to schedule, carbon taxes while others have implemented, or are planning to implement, cap-and-trade schemes. Carbon taxes set a price per unit of carbon emissions and let the market determine the associated quantity of emissions. Cap-and-trade systems determine a quantity of carbon emissions allowing the market to determine the price (World Bank, n.d. a). Either method generates government revenues for the jurisdiction in which the carbon pricing scheme is implemented that may be used to support social and climate justice.

To internalize the negative externalities associated with greenhouse gas emissions in a more equitable fashion, the proceeds from carbon pricing schemes could be used to attenuate the damaging climate impacts specifically on Black, Indigenous and other communities of color in the United States. An early example is provided by Fresno, California, where proceeds from a cap-and-trade program have been invested to fund new infrastructure in long neglected neighborhoods (Yale Climate Connections 2020). The consequences of extreme weather can include the loss of health, housing and employment, and are most difficult to manage for groups that have been, and continue to be, marginalized. A recent report by McIntosh et al. (2020) on the magnitude of the Black-White wealth gap in the United States complements earlier research by Daly et al. (2017) on the Black-White wage gap. The results highlight an urgent need to address existing racial and social injustice before climate change widens these gaps beyond magnitudes that are already unacceptable.

Regarding the question of what carbon price to set, the United States Environmental Protection Agency (US EPA) in 2017 recommended a tax of about $40 per metric ton of $CO_2$ with a gradual increase over time (US EPA 2017). In contrast, the results of Daniel et al. (2019) suggest that the tax should start above $100 as soon as possible and decline over time as uncertainty regarding the damaging impacts of climate change is resolved. In addition to disagreement over the amount and price paths of the tax, disagreement remains about how to allocate the proceeds, and how to adjust for domestic carbon prices at the borders. Given the uncertainty associated with the future course of climate change, it may be impossible to identify an optimal carbon pricing scheme at this time. Yet, the costs of postponing the decision may far exceed the benefits of greater certainty in the future. Agreement on any carbon price above zero now offers opportunities to make adjustments later on and appears to be the best available alternative. The timing of climate action critically determines the likelihood of its success, and carbon pricing is arguably the most powerful weapon in this fight because it offers simultaneous opportunities to mitigate social and climate injustice.

The Carbon Pricing Dashboard of the World Bank reports that, as of 2019, more than 20% of global greenhouse gas emissions were covered by carbon pricing schemes (World Bank, n.d. b). In the United States, some states along the East and West coasts have already begun to charge for greenhouse gas emissions as compensation for the damage they inflict on ecosystems, physical property, health and life by magnifying the risks from climate change. The Regional Greenhouse Gas Initiative (RGGI), a cooperative effort by ten states along the United States East Coast, supports the reduction of greenhouse gas emissions in the energy sector. The RGGI has generated a variety of benefits as summarized in a report entitled *Investment of RGGI Proceeds in 2017*, which was posted on the organization's web site in October 2019 (RGGI 2020). Among its successes, the RGGI counts improvements in energy efficiency along with the creation of jobs and worker training, direct bill assistance, greenhouse gas emissions reductions and public health benefits. The RGGI's past performance will likely encourage the inclusion of more states, sectors beyond energy and possibly greater emission cuts which would, undoubtedly, lead to an increase in the price of tradable allowances and further emission reductions. Faster and significant progress along these lines is, of course, necessary for a timely response to the climate crisis.

In a parallel effort, the Citizens' Climate Lobby, a non-profit nonpartisan grassroots organization in the United States, is advocating for the Energy Innovation and Carbon Dividend Act, which would place a fee on fossil fuels at their source and allocate the proceeds to households equally (Citizens' Climate Lobby, n.d.). Because high-income households tend to have larger carbon footprints than low-income households, the equal allocation of the collected revenue promises to, at least partially, counteract the regressive effects of a carbon tax (Rosenberg et al. 2018). The proposal outlines a national price on carbon that increases in a predictable fashion so that market participants have time to plan necessary resource re-allocations and investments in innovation. It also envisions carbon fees for imports at the border.

Despite a recognition by economists that carbon pricing schemes offer powerful tools in the global effort to reduce climate risks and build inclusive prosperity, current fossil fuel subsidies remain large around the world (Coady et al. 2019). These global subsidies, which favor resource allocations to high-carbon industries and create damaging price and quantity distortions (Erickson et al. 2020), should be phased out starting immediately. Successful subsidy removal, however, requires international coordination and cooperation among the governments within the G7 and G20, which critically depends on competent political leadership in numerous countries simultaneously. If governments were able to couple fossil fuel subsidy removal with the implementation of carbon pricing on a global scale, markets would be very effective at efficiently internalizing a large portion of the negative externalities from fossil fuel production and consumption.

While carbon pricing is widely recognized as both powerful and essential, it alone will not be able to transition the economy in time to prevent average global temperature increases with catastrophic consequences (Kramer 2019; Nuccitelli 2020). Government policies that encourage innovation in all sectors of the economy, cross-sector collaborations and public-private partnerships will be required to speed up the low-carbon transition both domestically and globally. Economic growth will have to occur with net-zero carbon emissions, and it must be equitable.

## 5.4   SUSTAINABLE RESOURCE ALLOCATIONS

Carbon pricing policies are important components of a just low-carbon transition, but even without them, businesses already have the tools that are needed for the required resource re-allocations. Resources are those assets, skills, systems and knowledge that are tied semi-permanently to the firm (Wernerfelt 1984). They are under the control of the firm and provide the foundation for the firm's strengths and weaknesses. These various resources may be viewed as forms of capital: financial, human, intercultural, environmental, atmospheric and so on. Budgeting techniques can be employed to make allocations of all forms of capital fairer and more sustainable. The basis for these improved resource allocations is the recognition that financial resources are not the only form of capital that requires a return. Atmospheric, environmental, natural, human, cultural and social resources can be depleted and must be assigned an opportunity cost in capital budgeting analyses to ensure they are employed in ways that are both productive and regenerative (Liesen et al. 2013).

Ignoring atmospheric capital in the resource allocation process threatens financial system stability as climate risks accelerate unpredictably, disrupt social and productive processes and impair asset values (NGFS 2019; Schellhorn 2018). A similar argument can be made when human, social and cultural forms of capital are ignored (Bourdieu 1984; Holt 1998). When Black and Indigenous people in the United States are paid less than White people with comparable qualifications, large parts of the population have inadequate access to healthcare and education, making them less productive, unable to participate in the economy and vulnerable in pandemics. As the COVID-19 pandemic has shown, we cannot afford to wait (White 2020). While social capital focuses on one's network, relationships and affiliations, cultural capital is described by Holt (1998) as "a set of socially rare and distinctive tastes, skills, knowledge, and practices" that are shared by participants within various market segments. When social and cultural forms of capital are depleted because people that possess this capital are not fairly compensated, firms experience a decline in revenue sources as their customer base narrows. Thus, resource allocations that reinforce existing social and climate injustice raise economic, social and political risks that can be destabilizing not only for individual firms and the financial system, but also for our political and public health systems. Opportunity costs for atmospheric capital may be approximated with available estimates for the social cost of carbon. Opportunity costs for social, cultural and other forms of capital are more difficult to estimate and are likely to vary across industries and firms, but they are clearly greater than zero. Owners of these forms of capital make valuable contributions to the functioning of our interconnected systems and the well-being of society. They must be fairly compensated in order to thrive.

Capital budgeting analyses by business managers that make explicit opportunity costs for non-financial forms of capital need support from investors who understand the importance of maintaining the functioning of interconnected social, political, economic, financial and ecological systems (TIIP n.d.). These investors view shareholder value creation as a small part of the much larger effort of creating value for the systems that constitute our society and support our lives on this planet (Hall and Hodgson 2018). A systems-oriented approach to investing requires a commitment to stewardship. Institutional investors' fiduciary duty includes engagement with legislators, regulators, managers and individual investors. In some cases, divestment may be the most impactful course of action. Collaboration on creating system value can be very effective as well as profitable, especially when employed by institutional investors with control over large asset portfolios (Naaraayanan et al. 2019). On racial equity, diversity and inclusion, philanthropy has proven to be an indispensable partner (Onek 2019). The financial sector is increasingly becoming aware of its unique responsibility and massive impact in shaping not only the economy, but also society and the natural environment on which our lives depend (Sullivan et al. 2019). Simultaneously, it is becoming clear that market participants will have to move much faster to mitigate the growing, interconnected climate and social risks.

## 5.5  PUBLIC-PRIVATE PARTNERSHIPS

Markets work best when stakeholders become interculturally competent and when the contributions of environmental and atmospheric capital are valued. Intercultural competence is "the ability to successfully communicate with people of other cultures" (Demangeot et al. 2013). In a healthy economy, those who possess intercultural competence leverage this resource to engage other individuals, firms, community groups, NGOs (Non-Governmental Organisations) and policy makers to reduce marketplace vulnerability. These stakeholders ought to receive a fair return on their contributions of the social and cultural capital necessary to develop and sustain a healthy economy. Just as important is the recognition that federal and municipal governments, as providers of public goods, make valuable contributions to society and require a commensurate return. In order to unleash the full capacity of markets to support social progress, it is necessary for the public and private sectors to collaborate and support one another in addressing society's enormous challenges.

Past contributions of the public sector have led to innovations that have been transformative in critical industries such as medicine, computers and energy (Mazzucato 2018). Yet, while much attention is paid to the size of government debt, the value contributions of public sector investments are rarely mentioned. Perkis (2020) points out that both federal and municipal debt together amounted to less than 70% of all private debt in the United States as of 2019. While the amount of spending by federal and municipal governments over the long term must not exceed reasonable limits, a more important question is what kinds of projects governments choose to fund. Supporting inclusive prosperity while rapidly reducing greenhouse gas emissions and the associated climate and health risks will require research and development and innovation on a scale that likely exceeds that of past public-private collaborations like the moon landing or the creation of the internet. Accomplishing these mutually dependent goals in a timely fashion will be impossible without effective government involvement.

Projects that would benefit from public-private partnerships include regenerative organic local agriculture that provides healthy, climate-friendly food also to disadvantaged communities and educates people on the preparation of these foods. Equally important are low-cost financing for community solar and wind energy, accessible and affordable public transportation, healthcare, as well as relevant education and professional training. These collaborative efforts could be at least partially funded with proceeds from the aforementioned removal of fossil fuel subsidies and carbon pricing. Public banks, which can be established as local green banks, as a national or international climate bank or as another bank created by a governmental body in the public interest, could help funnel state and municipal tax revenue to research, development and innovation that is both green and equitable in all sectors of the economy (Public Banking Institute 2020). In addition, both governments and businesses may tap into the growing market for green, social and sustainability bonds in order to signal to systems-minded investors the purpose of their fund allocations (Moody's Investors Service 2020 ).

The federal government or municipal governments could also make legal or regulatory changes to redirect fund flows that are currently contributing to social, racial and climate injustice. For instance, Dougal et al. (2018) find that racial hostility in some of the southern states causes Historically Black Colleges and Universities (HBCUs) to face higher financing costs for their bond issues compared to their non-HBCU peers. A possible remedy would be to offer tax benefits of holding these bonds, similar to those enjoyed by local investors, to investors in other states where these attitudes are less prevalent, provided those states exist. Removing legal barriers to funds flows across jurisdictions could help market arbitrage level the playing field in locations that are currently underserved.

Community-based organizations in the United States like UPROSE, as centers of intercultural competence, can be important facilitators of sustainable development in urban areas (UPROSE, n.d.). With a focus on climate justice advocacy, UPROSE is a Brooklyn, New York-based organization that promotes community engagement, sustainable justice and government accountability. NGOs such as UPROSE can push local governments toward greater transparency and accountability while fostering low-carbon and green economic growth that is coupled with social and racial justice. They may also initiate partnerships with worker-owned enterprises and cooperatives that are better suited to serve society than corporations with their dominant focus on shareholder wealth maximization. Human ingenuity is required as never before.

## 5.6 CONCLUDING REMARKS

This chapter argues that the dual goals of reducing the growing risks from climate change and supporting inclusive prosperity are interrelated and mutually dependent. To flourish, humanity must unleash its undivided power to counter existential threats, and use its talents to solve a multitude of diverse problems. Markets offer powerful mechanisms that can provide the scale required to efficiently build a healthy economy within a healthy society on a healthy planet. However, markets are likely to fail without the appropriate ELRF. Beyond that, markets' ability to deliver outcomes that are critical for our collective survival depend on the following system components: mechanisms that price greenhouse gas emissions and distribute the proceeds equitably, a widespread practice of including opportunity costs of non-financial forms of capital in budgeting processes, and public-private partnerships that enhance the efforts of both governments and markets in addressing our unprecedented global challenges. We are all in this together, and the time for action is now.

## REFERENCES

Bahn, K. 2020. "Clean Slate for Worker Power' Promotes a Fair and Inclusive Economy." Washington Center for Equitable Growth. https://equitablegrowth.org/clean-slate-for-worker-power-promotes-a-fair-and-inclusive-u-s-economy/

Bolton, P., M. Despres, L. A. Pereira Da Silva, F. Samama, and R. Svartzman. 2020. *The Green Swan*. Bank for International Settlements.

Bourdieu, P. 1984. *Distinction: A social critique of the judgement of taste*. Routledge.

CDC. 2019. "Climate Effects on Health." https://www.cdc.gov/climateandhealth/effects/default.htm

Citizens' Climate Lobby. n.d. "The Bipartisan Climate Solution" Retrieved from https://citizensclimatelobby.org/energy-innovation-and-carbon-dividend-act/

Coady, D., I. Parry, N.-P. Le, and B. Shang. 2019. "Global Fossil Fuel Subsidies Remain Large: An Update Based on Country-Level Estimates." IMF Working Paper WP19/89.

Daly, M. C., B. Hobijn, and J. H. Pedtke. 2017. "Disappointing Facts about the Black-White Wage Gap." *FRBSF Economic Letter*, September 5, 2017. Federal Reserve Bank of San Francisco.

Daniel, K. D., R. B. Litterman, and G. Wagner. 2019. "Declining $CO_2$ Price Paths." *Proceedings of the National Academy of Sciences* 116 (42): 20886–20891.

Demangeot, C., N. R. Adkins, R. D. Mueller, G. R. Henderson, N.S. Ferguson, J. M. Mandiberg, A. Roy, G. D. Johnson, E. Kipnis, C. Pullig, A. J. Broderick, and M. A. Zuniga. 2013. "Toward Intercultural Competency in Multicultural Marketplaces." *Journal of Public Policy and Marketing* 32: 156–164.

Dougal, C., P. Gao, W. J. Mayew, and C. A. Parsons. 2019. "What's in a (School) Name? Racial Discrimination in Higher Education Bond Markets." *Journal of Financial Economics*, 134 (3): 570–590. https://doi.org/10.1016/j.jfineco.2019.05.010

Dudley, W. C. 2011. "Financial Stability and Economic Growth." September 23, 2011. Federal Reserve Bank of New York. Available at https://www.newyorkfed.org/newsevents/speeches/2011/dud110923

Erickson, P., H. van Asselt, D. Koplow, M. Lazarus, P. Newell, N. Oreskes, and G. Supran. 2020. "Why Fossil Fuel Producer Subsidies matter." *Nature*. Available at https://www.nature.com/articles/s41586-019-1920-x.epdf

EPA. 2017. "The Social Cost of Carbon," Retrieved from https://19january2017snapshot.epa.gov/climatechange/social-cost-carbon_.html

Fender, I., M. McMorrow, V. Sahakyan, and O. Zulaica. 2019. "Green Bonds: The Reserve Management Perspective." *BIS Quarterly Review*. Available at https://www.bis.org/publ/qtrpdf/r_qt1909f.htm

GAO. 2018. https://www.gao.gov/products/GAO-18-188

Greve, Bent. 2017. How to measure social progress?. *Social Policy & Administration*, 51(7), 1002–1022.

Hall, M., and T. Hodgson. 2018. *Connecting the Dots: Understanding Purpose in the Investment Industry*. Thinking Ahead Institute. Willis Towers Watson. https://www.thinkingaheadinstitute.org/en/Library/Public/Research-and-Ideas/2018/10/Value-creation-connecting-the-dots

Holt, D. B. 1998. "Does cultural capital structure American consumption?" *Journal of Consumer Research* 25 (1), 1–25.

Honohan, P. 2019. "Should Monetary Policy Take Inequality and Climate Change into Account?" Working Paper 19–18, Peterson Institute for International Economics.

Islam, S. N., and J. Winkel. 2017. "Climate Change and Social Inequality." DESA Working Paper No. 152. United Nations, Department of Economic and Social Affairs.

Korten, D. 2010. "Ten Essential Principles for Healthy Markets" https://davidkorten.org/healthy-markets/

Krakoff, S. 2019. "Environmental Justice and the Possibilities for Environmental Law." *Environmental Law* 49: 229–247. U of Colorado Law Legal Studies Research Paper No. 19-19. https://ssrn.com/abstract=3407486

Kramer, D. 2019. "Should Carbon Emissions Be Taxed or Capped and Traded?" *Physics Today* 72 (12): 28–30.

Leiserowitz, A., E. Maibach, S. Rosenthal, J. Kotcher, P. Bergquist, M. Ballew, M. Goldberg, and A. Gustafson. 2019. "Climate Change in the American Mind: November 2019." Yale Program on Climate Change Communication. https://climatecommunication.yale.edu/publications/climate-change-in-the-american-mind-november-2019/2/

Liesen, A., F. Figge, and T. Hahn. 2013. "Net Present Sustainable Value: A New Approach to Sustainable Investment Appraisal." *Strategic Change* 22 (3-4): 175–189.

Lewis, Renee. 2014. "Life in Michigan's Dirtiest ZIP Code," *AlJazeera America*. http://america.aljazeera.com/articles/2014/3/3/michigan-tar-sandsindustryaccusedofacting-withimpunity.html

Maslow, A. H. 1943. "A theory of human motivation." *Psychological Review* 50: 370–396.

Mazzucato, M. 2018. "Who Really Creates Value in an Economy?" *Project Syndicate*. https://www.project-syndicate.org/commentary/economy-value-private-public-investment-by-mariana-mazzucato-2018-09

McIntosh, K., E. Moss, R. Nunn, and J. Shambaugh. 2020. "Examining the Black-White Wealth Gap." The Hamilton Project, Brookings.

Moody's Investors Service. 2020. *Green, Social and Sustainability Bond Issuance to Hit Record $400 Billion in 2020*. Moody's. Report Number 1210908.

NASA. n.d. https://climate.nasa.gov/scientific-consensus/

Naaraayanan, S. L., K. Sachdeva, and V. Sharma. 2019. "The Real Effects of Environmental Activist Investing." https://ssrn.com/abstract=3483692

NGFS. 2019. *A Call for Action: Climate Change as a Source of Financial Risk.* NGFS first comprehensive report. https://www.banque-france.fr/sites/default/files/media/2019/04/17/ngfs_first_comprehensive_report_-_17042019_0.pdf

Nuccitelli, D. 2020. "With the EN-ROADS Climate Simulator, You Can Build Your Own Solutions to Global Warming," *Yale Climate Connections.* January 24, 2020. https://www.yaleclimateconnections.org/2020/01/en-roads-climate-solutions-simulator/

OECD. 2015. *In It Together: Why Less Inequality Benefits All,* OECD Publishing, Paris. http://dx.doi.org/10.1787/9789264235120-en

Onek, M. 2019. "Impact Investing and Racial Equity: Foundations Leading the Way," *Stanford Social Innovation Review.* February 25, 2019. https://ssir.org/articles/entry/impact_investing_and_racial_equity_foundations_leading_the_way

Patterson, J. 2019. *Fossil Fueled Foolery,* National Association for the Advancement of Colored People. https://live-naacp-site.pantheonsite.io/wp-content/uploads/2019/04/Fossil-Fueled-Foolery-An-Illustrated-Primer-on-the-Top-10-Manipulation-Tactics-of-the-Fossil-Fuel-Industry-FINAL-1.pdf

Penn, Ivan (January 5, 2020) "N.A.A.C.P. Tells Local Chapters: Don't Let Energy Industry Manipulate You." *New York Times* retrieved from https://www.nytimes.com/2020/01/05/business/energy-environment/naacp-utility-donations.html

Perkis, D. F. 2020. "Making Sense of Private Debt." *Page One Economics.* Federal Reserve Bank of St. Louis. https://research.stlouisfed.org/publications/page1-econ/2020/03/02/making-sense-of-private-debt/

Podesta, J. 2019. "The Climate Crisis, Migration, and Refugees." *Brookings Blum Roundtable on Global Poverty.* https://www.brookings.edu/research/the-climate-crisis-migration-and-refugees/

Public Banking Institute. 2020. "How Germany's Public Bank Will Keep the Economy Afloat During the Coronavirus Outbreak." https://www.publicbankinginstitute.org/2020/03/19/how-germanys-public-bank-will-keep-the-economy-afloat-during-the-coronavirus-outbreak/

Raworth, K. 2017. *Doughut Economics: Seven Ways to Think like a 21st Century Economist.* White River Junction, Vermont: Chelsea Green Publishing.

RGGI. 2020. "Investments of Proceeds" Retrieved from https://www.rggi.org/investments/proceeds-investments

Rockström, J., W. Steffen, K. Noone, Å. Persson, F. S. Chapin, III, E. Lambin, T. M. Lenton, M. Scheffer, C. Folke, H. Schellnhuber, B. Nykvist, C. A. De Wit, T. Hughes, S. van der Leeuw, H. Rodhe, S. Sörlin, P. K. Snyder, R. Costanza, U. Svedin, M. Falkenmark, L. Karlberg, R. W. Corell, V. J. Fabry, J. Hansen, B. Walker, D. Liverman, K. Richardson, P. Crutzen, and J. Foley. 2009. Planetary Boundaries: Exploring the Safe Operating Space for Humanity. *Ecology and Society* 14(2): 32. http://www.ecologyandsociety.org/vol14/iss2/art32/

Rosenberg, J., E. Toder, and C. Lu. 2018. *Distributional Implications of a Carbon Tax.* Urban-Brookings Tax Policy Center. https://energypolicy.columbia.edu/research/report/distributional-implications-carbon-tax

Schellhorn, C. 2018. "The Low-Carbon Transition and Financial System Stability." In S. Boubaker, D. Cumming, and D. K. Nguyen (Eds.), *Research Handbook of Finance and Sustainability.* Edward Elgar Publishing. 410–420.

Schellhorn, C. 2020. "Financial System Stability, the Timing of Climate Change Action and the Federal Reserve." *Journal of Central Banking Theory and Practice* 9 (3): 45–59.

SEC. 2010. https://www.sec.gov/rules/interp/2010/33-9106.pdf

Sullivan, R., W. Martindale, E. Feller, M. Pirovska, and R. Elliott. 2019. *Fiduciary Duty in the 21st Century.* UNPRI. https://www.fiduciaryduty21.org/publications.html

TIIP. n.d. "System-Level Investing: Identifying, Measuring, and Managing Environmental, Social, and Financial System Risks and Investing in Solutions to Systemic Problems" Retrieved from https://www.tiiproject.com/system-level-investing/

UPROSE. n.d. "Who We Are," Retrieved from https://www.uprose.org/mission

Wernerfelt, B., 1984. "A Resource-Based View of the Firm." *Strategic Management Journal*, 5(2), 171–180.

White, Amanda. March 31, 2020. "Coronavirus, Climate Change Parallels." https://www.top1000funds.com/2020/03/risk/?amp

World Bank. n.d.a. "Pricing Carbon." https://www.worldbank.org/en/programs/pricing-carbon

World Bank. n.d.b. "Carbon Pricing Dashboard." https://carbonpricingdashboard.worldbank.org/

Yale Climate Connections. January 3, 2020. "California's Cap-and-Trade Program Is Paying for Improvements in Disadvantaged Communities." https://www.yaleclimateconnections.org/2020/01/californias-cap-and-trade-program-is-paying-for-improvements-in-disadvantaged-communities/

# 6 Green Banking
## An Environmental Shield for Sustainable Growth in India

*Samrat Roy and Xavier Savarimuthu, SJ*

## CONTENTS

## 6.1   INTRODUCTION: BACKGROUND AND DRIVING FORCES

Global climate change has become internationally recognized as a serious challenge over the last few decades. Adverse consumption and production externalities have drastically affected the ecological balance. The crisis has been further aggravated due to the overuse of natural resources, rapid urbanization, deforestation, water pollution, and shrinking energy capacities. A globalized economy does not exist in isolation – any tremor felt in one corner of the world will shake the base of the global economy. However, it cannot be denied that development also means inclusiveness, which entails a greater participation of the masses with active involvement in the decision making process. The inclusiveness is consistent with the pro-market orientation of the larger segment of the society, but this engagement does not advocate for growth at the cost of the environment. The concern for sustainability stems from the nexus between the needs of the environment, economy, and society, which are the three pillars of the sustainability tripod.

Money has been a part of human history for more than 3,000 years, beginning with a barter system initially used for dealing in goods and services, until there was an evolution of money to encompass metal coins, paper notes, gold, silver, and others. In a monetary system, the goods or services are assigned a value in units that the customer uses to pay for it. Now, in the 21st century, there has been further evolution of the monetary system. Instead of paying for a commodity in metal coins or paper notes, the consumer has a better and easier way to pay for the goods or services. This new, disruptive means of payment is digital payment. In simple terms, digital payments involve payment for the commodity by electronic means without the use of cash or a cheque. In this system, the sender and the receiver use digital means for transactions.

Digital payments are slowly overtaking the financial transactions market, which has traditionally relied on cash and cheques. Sweden and other Scandinavian countries are on the verge of becoming completely cashless economies. Digital transactions have transformed the mode of payment and changed the retail and banking sector of the world's economies.

This growth in India is because of the success of Unified Payment Interface (UPI), an electronic-based payment system that has drastically changed the payment ecosystem in India. The initiatives undertaken by the government and the Indian Federal Reserve, the Reserve Bank of India, played a major role in bringing about this significant change in the Indian monetary system.

The development of the online payment system began in the early 1990s. In 1994, Stanford Federal Credit Union was the first institution in the world to provide an online payment service. In India, BILLDESK, launched in 2000, was the first payment aggregator. The e-commerce giant Flipkart was launched in India in 2007 and now holds nearly 40% of the market share. The growth of the e-commerce business has increased the need for a reliable payment gateway to facilitate online shopping. To support livelihoods from across different strata of society, any emerging economy needs to have a sound financial and banking structure. The concern regarding environmental protection was reflected in the advent of Automated Teller Machines (ATMs) as recently as 2001, as far as technological innovations in the Indian banking sector were concerned. Policy makers stressed quality indicators for strengthening the efficiency of the banking sector in India. The adoption of technological innovation helped to upgrade the quality of banking services. The banking sector gained momentum, offering an array of online services to its customers. Policy makers have coined the term "Green Banking," which represents an endeavor to simultaneously protect the environment and satisfy customers. The micro benefits accrue to firms and households, whereas macro benefits accrue to the sustainable growth of the nation. India's digital payment systems open up new horizons in green banking. The positive externalities of green banking or a stride toward a paperless economy was highly pronounced in information technology (IT) enabled services. E-commerce was largely felt across fast-moving sub-sectors in services such as in the business, communications, transport, and tourism sectors.

The digital payment market is in a boom period, as there are continuous improvements in technologies that make these services easy and simple to use, while also spreading awareness of various methods of digital payments through different electronic mediums.

There are multiple benefits to using digital transactions:

1. Digital banking offers transparency, accountability, and easy accessibility. Transactions can be carried out at any time on any day throughout the year.
2. It is cost effective. Most digital payment services such as UPI do not charge service fees or transaction fees. Other digital payment services are progressing toward reducing costs.

3. Rewards and discounts are offered to customers who use digital payment apps and mobile wallets. These perks serve as motivation to encourage using digital payment services.
4. Digital banking provides an electronic record of transactions – all transaction records can be kept digitally. Customers can keep track of each and every transaction.
5. It can keep corruption under control. The digitization of transactions helps the government keep detailed records, and eliminates the circulation of black money and fake notes in the economy. It also reduces the cost of minting currency.

## 6.2 LITERATURE REVIEW

A study by Vally and Divya (2018) focused on how government initiatives like Digital India, combined with the increasing use of mobile phones and more widely available internet, have resulted in the transformation from a traditional cash payment system to the modern digital payment system in India. The aim of this chapter is to review the positive impact of digitizing the payment system. This chapter also assesses how well customers embrace the new digital payment services.

Kabir et al. (2015) analyzed the efficiency, convenience, and timeliness of a digital payment system that is being adopted into the financial systems of economically-developed and economically developing countries with an aim to simplify payments during business transactions. This chapter recognizes the patterns of previous findings, and highlights and recommends key areas for future research. Singh and Rana (2017) explored how the increase in the use of the internet and mobile phones has boosted the digitization of India's payment system, based on a questionnaire to collect responses. From these responses, the researchers found that there was no great variance in consumers' perceptions based on factors like age, gender, and income. However, education seemed to significantly influence how the consumers embraced the digital payment system. Fatonah et al. (2018) reported that various technology innovations have led to a shift from cash-based transactions to electronic-based transactions with a digital payment system serving as a better alternative to both cash and trade-barter payment systems. This study also analyzes the e-payment systems with regard to e-commerce. Khan et al. (2017) discussed the services provided by the new digital payment systems with a variety of online payment methods, and demonstrated that this form of payment is increasingly popular. This chapter analyzes the current status, and growth, of e-payment systems in India. This study evaluates various factors that influence customers adopting e-payment systems, and examines the effect of green banking initiatives on financial performance, while also focusing on the profitability of banks practicing green banking.

## 6.3 OBJECTIVES OF THE STUDY

The objective of the current study is to examine the effect of green banking initiatives on the financial performance of banks in India. This incorporates factors related to profitability, net income, and expense of the organization.

## 6.4   DATA AND METHODOLOGY

A panel structure is constructed to consider the impact of income and expenses on the levels of profitability during the green banking era. The annual data were collected for a period from March 2015 to March 2019. The sampling frame comprises selected private and public sector banks. Two public banks, the State Bank of India and Punjab National Bank, and two private banks, ICICI Bank and HSBC Bank, were selected for our study. All of these banks have implemented green banking initiatives since 2015. The study conducted by Nanda and Bihari (2012) corroborates this methodology.

The empirical specification is:

$$Y_{it} = \beta_o + \beta_1 NetInc_{it} + \beta_2 Expenses_{it} + \beta_3 GreenBanking_{it} + error\ term \quad (6.1)$$

Here,

$Y_{it}$ = Income margin of bank $i$ in year $t$.
$NetInc_{ab}$ = Net Income of the $i$th bank in year $t$.
$Expenses_{ab}$ = Expenses incurred by the $i$th bank in year $t$.
$GreenBanking_{ab}$ = Adoption of green banking initiatives by the bank $i$ in year $t$.

The slope coefficients are given by $\beta_1$, $\beta_2$, and $\beta_3$.

The data is analyzed using the panel regression method. The income margin is the dependent variable.

## 6.5   EMPIRICAL FINDINGS AND INTERPRETATIONS

The estimated equation is:

$$Y_{it} = 16.96 + 0.007\ NetInc_{it} - 0.008\ Expenses_{it} + 0.33\ GreenBanking_{it} \quad (6.2)$$

p-values (0.00)(0.02)(0.03)(0.02).

The findings confirm that for the Indian banking sector, there exists a significant relationship between net income and expenses for green banking with income profitability during green banking implementation. If the net income rises by one unit, then the income margin of the selected banks will rise by 0.007 units. This also calls for mitigating non-performing assets.

The expenses will significantly decline with the implementation of green banking. This finding infers that the adoption of green banking practices have led to cost reduction. The dummy variable for green banking implementation appears to be statistically significant, justifying the inclusion of this variable.

## 6.6   CONCLUSIONS AND RECOMMENDATIONS

This study empirically explores the relationship between banking profitability and the adoption of a green banking method. The findings of this study confirm that all the explanatory variables significantly contribute to the profitability of the Indian banking sector. The study can be further enriched by conducting a cost-benefit analysis of implementing green banking. The measurement criteria regarding green banking can then be analyzed.

There are multiple components that the Indian banking sector should focus on:

1. Sustainability issues require awareness, achieved through using international guidelines and frameworks
2. Reporting on sustainability with formal frameworks and clear policies, which is urgently needed
3. Human capital development, which is urgent for banking personnel
4. Altering the management structure to focus on a skill-based knowledge economy
5. Disseminating information
6. Banks focusing on CSR areas

International endeavors achieved through funding provided by the International Monetary Fund and the World Bank will be instrumental in accelerating green banking initiatives in economically developing countries. The study indicates that green banking initiatives have been instrumental in channelizing banking prospects, as well as sustainable development for the near future. The returns from such investments will help in reducing carbon footprints, creating diverse forms of electronic banking, achieving environmental consciousness, and financing more environmentally friendly projects. Technological innovations will motivate bankers to cater to a larger section of society using methods that are consistent with green banking for a low carbon, sustainable economy.

## REFERENCES

Fatonah, S., A. Yulandari, and Ferry Wahyu Wibowo. 2018. "A Review of E-Payment System in E-Commerce." *Journal of Physics* 1140 (1): 1742–6596.

Kabir, Muhammad Auwal, Siti Zabedah Saidin, and Aidi Ahmi. 2015. "Adoption of e-Payment Systems: A Review of Literature." Presented at the International Conference on E-Commerce in Kuching, Sarawak. https://www.researchgate.net/publication/303329794.

Khan, Burhan UI Islam, F. Rashidah, Asifa Mehraj, Adil Ahmad, and Shahul Assad. 2017. "A Compendious Study of Online Payment Systems: Past Developments, Present Impact, and Future Considerations." *International Journal of Advanced Computer Science and Applications* 8 (5): 256–271.

Nanda, Sibabrata, and Suresh Chandra Bihari. 2012. "Profitability in Banks of India: An Impact Study of Implementation of Green Banking." *International Journal of Green Economics* 6 (3): 217–225.

Singh, Shamsher, and Ravish Rana. 2017. "Study of Consumer Perception of Digital Payment Mode." *Journal of Internet Banking and Commerce* 22 (3): 1–14.

Vallys, K. Suma, and K. H. Divya. 2018. "A Study on Digital Payments in India with Perspective of Consumer's Adoption." *International Journal of Pure and Applied Mathematics* 119 (15): 1259–1267.

# 7 Green Computing
## An Eco-Friendly and Energy-Efficient Computing to Minimize E-Waste Generation

*Siddhartha Roy*

## CONTENTS

## 7.1 INTRODUCTION

Green computing is the study and practice of reducing computers' use of resources and consequent environmental effects. This emerging concept is used to reduce harmful substances and save the environment from the damaging impacts of various computer hardware, such as central processing units (CPUs), monitors, and other electronic devices.

In broader terms, green computing reduces ecological impact through a systematic study of designing, manufacturing and engineering, and disposing of computing devices. The importance of green computing is gaining recognition among not only environmental organizations but also from other industrial sectors. In 1992, the United States Environmental Protection Agency first promoted "Energy Star," an eco-friendly and energy-efficient equipment control to initiate green computing practices (Wikipedia, n.d.). Concurrently, the Swedish organization Tjänstemännens Centralorganisation (TCO) Development launched the TCO program to monitor electrical emissions from Cathod Ray Tube (CRT)-based computer monitors (Boivie 2008). Green computing attempts to achieve economic viability and improve system performance in a green manner. To encourage the

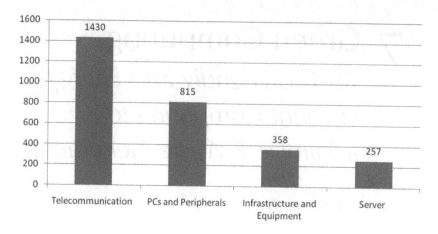

**FIGURE 7.1**   $CO_2$ emissions in the year. (Source: Statista Research Department, 2020.)

recycling of computing devices and reducing the use of various toxic materials, computer manufacturing companies have sought investment funding. In recent years computer manufacturing companies such as Apple, Microsoft, HP, and Dell have taken an initiative to embrace green business practices in terms of cost reduction and finding more eco-friendly alternatives for computing devices (Greig 2018). There are three key practices of green computing: developing environmentally sustainable products, designing energy-efficient hardware components, and improving disposal and recycling procedures. Green manufacturing (Schipper and de Haan 2005), which is used to minimize waste during the manufacture of computing devices, and green disposal, i.e., recycling superfluous electronic equipment and other subsystems represent important initiatives.. At present, internet and communication technology (ICT) contributes to 5%–9% of the world's total energy consumption and is expected to consume 20% by the year 2030 (Enerdata 2020). Figure 7.1 illustrates the total $CO_2$ emissions due to telecommunications and information technology (IT) in the year 2020.

Within the next ten years, it is estimated that billions of IT enabled devices will produce 3.5% of global emissions of $CO_2$ and reach 14% by the year 2040 (The Guardian 2017). This chapter assesses various technological solutions, such as green data centers, power optimization, virtualization, and cloud computing identifies key issues, and evaluates various technological approaches along with delineating the important steps that have been initiated toward green computing.

## 7.2   PREVIOUS WORKS

The United States Environmental Protection Agency's "Energy Star" designation represented the first step toward green computing and was primarily used to monitor the energy consumption of computing equipment and electronic devices. IBM Corporation offered eco-friendly computing devices that consume less power. This move reduced the carbon dioxide emissions by approximately 3 million metric tons and saved about 4.6 billion kWh in electricity (Clancy 2012). More recently, Google established a built-from-the-ground-up data center in the Dalles, Oregon area. The 30-acre site was chosen for cheap electricity from the 1.8 million kilowatt hydroelectric dam as well as cheap land and a welcoming local government (Wheeland 2007). Green computing research has focused on reducing energy and electricity consumption in data centers, designing eco-friendly computing devices which have minimal impact on environment as various computing devices generate e-waste and finally deployment of cloud computing for less use of hardware, software, and server. Pahlevan et al. (2017) presented an optimization framework for managing green data centers using multilevel energy reduction techniques that save up to 96% electricity bills. Riyaz et al. (2009) have described the impact of ICT and green computing on the environment (Schipper and de Haan 2005). They measured energy consumption in data centers, its efficiencies, and its virtualization impact. Sinha et al. (2011) proposed a dynamic threshold-based approach for reducing power consumption at data centers by live migration of the virtual machines and switching off idle machines. Garg and Buyya (2012) have discussed designing of environment-friendly computing, storage devices, and networking and communications systems with the main emphasis on recyclability, reducing electronic waste. Worthington and Brooks have focused on green technology strategies for computers and telecommunications, with robust recycling and reuse of equipment and components, along with green IT strategies and policies as sustainable business practices. (Worthington 2017; Brooks et al. 2010). Designing cloud computing framework, basic architecture, Software as a Service (SaaS) have also been a focus of recent research (e.g., Nair and Gopalakrishna 2009; Anandharajan and Bhagyaveni 2011; Taufiq-Hail et al. 2017).

## 7.3 DIFFERENT APPROACHES TO GREEN COMPUTING

Technological progress has had a significant and worldwide negative impact in the form of climate and other environmental changes, which are detrimental for the survival of humanity. Globally, the importance of using natural resources and reducing power consumption has been realized, an understanding that will protect present and future generations. It has been predicted that, if drastic steps are not taken to design eco-friendly computers, computer chips will require more electricity than our entire global energy production by 2040 (Dockrill 2016). Consequently, it is very important to adopt green computing to reduce power consumption and develop a sustainable framework to reduce carbon dioxide generation. Approaches toward green computing include virtualization, which uses software modifications that are designed to simulate hardware to reduce computers' consumption of power. Another approach is cloud computing, an on-demand computing service that allows servers, software, networks, or any other information to be accessible through the internet, now commonly used in the IT industry in many countries. Cloud computing also reduces the use of infrastructures (hardware, software, and other platforms supplied by the service provider based on demand), which is in turn reduces overall energy consumption (Kaur and Chana 2015).

### 7.3.1 VIRTUALIZATION

Currently, a computer uses almost 600 kWh and emits 175 kg of $CO_2$ per year, assuming it runs for 8 hours per day (Energuide, n.d.). Related to the growth of IT infrastructures, energy and power costs have also increased. Developing new hardware and minimizing everyday expenditure is a high priority for the IT industry (Riyaz et al. 2010). Virtualization, where a large number of servers and equipment are connected together to give a better output, has been adopted by various companies in recent years (Kurp 2008). Through virtualization, one set of physical hardware is used to simultaneously run multiple logical computer systems (Kusnetzky 2007). The primary objective of virtualization is to efficiently use available system resources (Kurp 2008). Virtualization also maximizes the CPU processing power on other servers. In the early 1960s, IBM first launched the concept of virtualization using their mainframe operating systems. In the late 1990s, this concept was commercialized for "x86"-compatible computers (Karthikeyan et al. 2019). With virtualization, several physical systems could be combined into one single virtual machine, thereby reducing the use of the original hardware, which lowered power and cooling energy consumption. Intel Corporation facilitated virtualization through proprietary virtualization that was compatible with the "x86" instruction set into product lines for each of their CPUs. Storage virtualization is used to store data just once instead of storing copies on a number of storage devices (Automation Business Technologies 2011). This means that copies of data can be stored once in the shared storage subsystem instead of on every computer's disks. Thus, virtualization reduces the number of storage devices, as well as the hardware components. In turn, power consumption, operational costs, and administrative costs of back up data are reduced. Figure 7.2 shows the server virtualization where multiple virtual servers work simultaneously on one physical server.

### 7.3.2 CLOUD COMPUTING

Cloud computing, another important approach toward green computing, refers to a situation where servers, storage, and applications can be accessed through the internet.

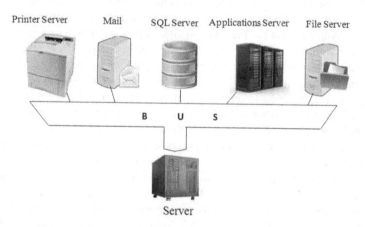

**FIGURE 7.2**   Server virtualization architecture.

According to the McKinsey report, IT companies produce nearly 1 gigaton of carbon dioxide ($CO_2$) emissions a year, which is 2% of total global $CO_2$ emissions (Creyts et al. 2007). The world's rapidly increasing demand for computation and data storage accounts for 1.54 gigatons of $CO_2$ emissions in 2020, or 3% of global emissions. Since the ever-increasing demand for computing devices cannot be reduced, adopting cloud computing is an effective approach to deliver computing in a more efficient, environmentally friendly way. Cloud computing can improve operational efficiency and decrease capital expenditure on IT resources by reducing energy consumption and carbon emissions (Jain et al. 2013). The key technology necessary for energy-efficient clouds is virtualization which has already been discussed. Two giant companies, Microsoft and Google, have already initiated to build energy savings cloud computing in the industry. It has been shown in the study conducted by Microsoft that companies adopting cloud computing can reduce energy consumption up to 90% than those less efficient companies running their application using traditional client server-based models (Kaur and Singh 2013). Google, one of the pioneers of modern-day cloud computing, has adopted a completely different computing model (Priya et al. 2013). Google produces cloud-based product platforms, such as Gmail and Google Docs, through a web browser to reduce the costs for purchasing software licenses and maintaining databases. More than two million businesses have signed up for Google's cloud-based products, and the adoption rate is increasing (Fe Bureau 2018).

There are many factors associated with cloud computing that reduce various IT industries' use of energy and production of carbon emissions, including effective use of servers, multi-tenancy, and data center efficiency. Cloud providers radically improve the power usage effectiveness (PUE) of their data centers (Yuventi and Mehdizadeh 2013).

Through storage as a service (SaaS), multiple companies use the same infrastructure and software, reducing the overall energy use and associated carbon emissions. Cloud providers monitor and predict the demand, and they allocate the resources dynamically, as per the organization's requirements (Prasad et al. 2019). Figure 7.3 shows a cloud-based system.

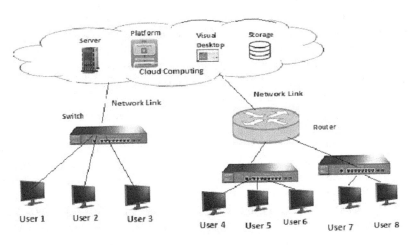

**FIGURE 7.3**   Cloud computing architecture.

Businesses are rapidly moving from the traditional client-server-based system to cloud systems because of the scale and pay-per-use facility of cloud infrastructure that simultaneously provide energy and resource effectiveness (Yamini 2012).

### 7.3.3 E-Waste

Green computing aims to minimize the environmental hazards of materials associated with computing devices, in particular by reducing toxic waste. This can be achieved by recycling electronic components and using fewer toxic substances during manufacturing. Computer circuit boards contain mercury, lead, cadmium, arsenic, beryllium, and other toxic chemicals. These materials are found throughout electronic devices: beryllium is used in motherboards and connectors; chromium and mercury are used in various relay devices such as routers and switches; cadmium is generally found in chip resistors, semiconductors, cables, and wires; lead oxide and cadmium are used in cathode ray tube monitors; mercury is used in flat screen monitors; and polychlorinated biphenyls are used in transformers and batteries (Lundgren 2012).

These materials are toxic and cause serious health effects in humans (Riaz et al., 2009). At present, roughly 120 million tablets, 350 million personal computers, and 270 million smartphones are sold worldwide every year (Garg and Buyya 2012). The use of electronic gadgets during the past two decades has been very high. E-waste is polluting the soil, water, and environment, leading to concerns around pollution control and environmental safety. Various toxic metals like lead and cadmium accumulate due to the gradual build-up of non-biodegradable e-waste. E-waste contains harmful materials that require special handling and recycling. There are three steps for recycling harmful materials: detoxification, shredding, and refining (Jofre and Morioka 2005). In the first step, detoxification, the various toxic components used in computing devices, such as lead, mercury, cadmium, beryllium, are identified and removed. During the second step, shredding is used to further separate the hazardous materials. The third step yields the raw materials, which have a minimal impact on the environment. A common e-waste practice to decrease the adverse effects of electronic equipment is the 3R process: reduce, recycle, and reuse. We should emphasize the reuse and recycling processes rather than realistically anticipate any decrease in the use of electronic equipment. Different companies such as Dell, HCL, and Microsoft are now adopting eco-friendly alternatives for industrialization and sustainable development (Begum and Nadaf 2014).

Isolating toxic substances through systematic planning can reduce adverse environmental effects. Unwanted and potentially hazardous electronic equipment and toxic metals in computing devices should be handled in a convenient and environmentally responsible manner (Annual Report. 2006, June 2007, Greenpeace International). E-waste can become a valuable source if it is properly used and reused. Waste Electrical and Electronic Equipment Directive (WEEE) or the European Community Directive 2012/19/EU on waste electrical and electronic equipment along with the RoHS (restriction of the use of certain hazardous substances in electrical and electronic equipment) Directive 2011/65/EU, which together became European Law in February 2003, have a huge impact on green computing (European Commission, n.d.). Presently, Dell, Cisco, and Apple are few examples of manufacturers who are developing less toxic, biodegradable, and eco-friendly computing equipment around the world.

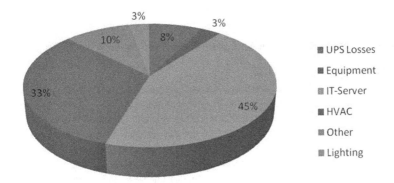

**FIGURE 7.4**    Typical breakdown of data center energy (Tiwari 2011).

### 7.3.4 GREEN DATA CENTERS

Data centers are the most complex and energy-demanding centralized computing and network equipment due to high internal loads and uninterrupted services
(Balaras et al. 2017).

Figure 7.4 shows a data center's typical energy use for various components.

Green data centers are being built to reduce carbon dioxide emissions and minimize heat dissipation while creating energy-efficient and environment-friendly
devices (Jaureguialzo 2011). Green data centers use green electronic devices and
components and software techniques to optimize power consumption (Gao 2014).
Various parameters, such as the use of power, energy, and carbon, are regularly
monitored. Power optimization coding algorithms are used to reduce the strain on
hardware and other resources. Examples of such algorithms are stack reduction
algorithm (SRA), integer-bit power allocation algorithm, and computation and transmission rate-based algorithm (CTRB). Several energy metrics have been proposed
to develop green data centers (Wang and Khan 2011). Among the most powerful
metrics are power usage effectiveness (PUE), data center infrastructure efficiency
(DCIE), data center performance efficiency (DCPE), and carbon usage effectiveness
(CUE). These metrics are used to measure the "greenness" of an existing data center (Sego et al. 2012). Table 7.1 contains various energy metrics, along with a brief
description of them.

Globally, the green data center market is expected to grow from USD 37.9 billion
to USD 197.72 by 2023, as shown in Figure 7.5.

## 7.4   GREEN COMPUTING INITIATIVE FROM
## AN INDIAN PERSPECTIVE

With more than 20,000 IT companies in India, India is an IT superpower. The
energy use in the IT industry is growing in an exponential manner. Data centers
are some of the greatest energy consumers in the industrial sector. Presently, India
generates 5 billion 200 million kg of e-waste (Ramesh and Joseph 2006). Figure 7.6

## TABLE 7.1
## Various Energy Metrics Used in Green Data Centers

| Energy Metrics | Description |
| --- | --- |
| PUE | PUE (power usage effectiveness) is defined by total power used by the data center building/power used of IT equipment |
| DCIE | DCIE (data center infrastructure efficiency) is defined by 1/PUE. Based on this measurement, corrective measures can be taken to implement a green data center. DCIE also defined by DCIE = (IT equipment power /total facility power) × 100% |
| DCPE | It is a natural evolution from PUE and DCE and is computed as DCPE = useful work/total facility power |
| CUE | It measures the carbon intensity of IT deployment. It measures the carbon gas emitted from the data center on a daily basis. It can be computed as CUE = total $CO_2$ emissions from DC energy/total IT energy |

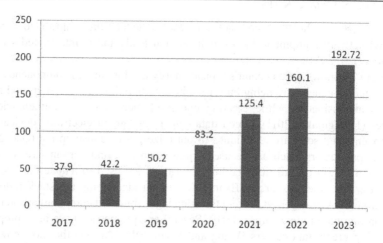

**FIGURE 7.5**   Global green data center market trend. (Source: Pune, India, March 27, 2019/ Marketers Media.)

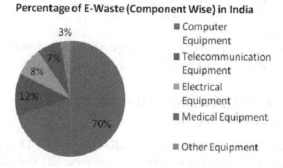

**FIGURE 7.6**   Pie chart for component wise e-waste percentage in India. (Source: www. economictimes.indiatimes.com/tech/hardware, November 5, 2012.)

shows e-waste generation from different sectors, based on the Assocham-cKinetics (2016) study.

Due to the exponential growth of the Indian IT industry, the e-waste generated by the IT infrastructure is significant. The informal sector generates approximately 95% of e-waste (Park and Soni 2019). The lack of awareness, including at the corporate level, is a major challenge in India. Moreover, a large number of people in economically developing countries express little concern for the effect of the toxic materials used in the computing devices and IT-associated equipment like motherboard and monitor. The Indian government should impose strict legislation to require companies to maintain the green computing standard. Additionally, the government should initiate a special awareness program in all sectors to teach organizations about reducing energy consumption, which minimizes the electricity bill, and using nontoxic materials in equipment to make the organization maintain occupational health and safety. As of now, the government has 312 registered recycling facilities across 19 states that have the capacity to recycle 0.78 million metric tons (Sheth 2019).

When the Indian IT industry takes the initiative to achieve green computing, there will be a remarkable change in this sector's sustainability (Ramesh and Joseph 2006).

The following section will discuss some Indian giant IT companies' initiatives toward green computing.

Wipro Infotech, a leading IT provider and business transformation service, has developed an eco-friendly commitment (Wipro 2010). Wipro manufactures greenware desktops that are completely free from carcinogenic chemicals, such as brominated flame retardants (BFRs) and polyvinyl chloride (PVC) (Nadaf and Nadaf 2014). These toxic substances' absence ensures that recycling these electronic products is easier; thus, launching PVC-free and BFR-free products is a major breakthrough in Wipro Infotech's recycling policy. Producing these toxin-free materials makes Wipro a first in India and among very few companies globally in this area. AT&T has announced its entrance into cloud computing, which reduces the burden to build and maintain IT infrastructures and services (Weissberger 2019). The multinational companies Hewlett-Packard (HP), Dell, and Acer have all adopted "green computing" in India (Jena 2013).

To reduce e-waste, these companies have launched energy-efficient computers made from recycled materials. HP has created a notebook battery replacement program in India. Dell and HP have started a recycling program through safely removing scrap components to reuse materials in an environment friendly manner, in addition to working on sustainable products. They have launched a recycling program in India to develop an energy-efficient paper laptop from recyclable materials. Repairing recyclable laptops is relatively easier than traditional laptops. Presently, various companies have developed ultra-high shearing (UHS) technology, which is based on the principle of ultra shearing, and it does not use any chemical additives (Panda 2013).

## 7.5  ISSUES AND FUTURE DIRECTION OF GREEN COMPUTING

Green computing has a great impact on environmental sustainability, but there are some issues when implementing green computing. First, the research cost is very high, and most eco-friendly devices are presently very costly as a consequence replacing existing devices with eco-friendly devices increases expenditures. Second, the lack of enforcing policy standards and developing Green Maturity Models (GMM) (Nanath and Pillai 2017) for IT infrastructure are major issues for environmental sustainability. Main focus of green computing is to decrease the cost of energy at data centers, in turn, reducing generated carbon. The next focus of green computing is innovation and improving overall corporate social responsibility efforts based on developing computing strategies (Kazandjieva et al. 2011). In the future, the main criteria for green computing will be efficiency rather than reducing energy consumption (Soomro and Sarwar 2012). A certificate of green computing should be mandatory for all organizations that manufacture computing devices. Computing devices' resource use can be achieved through the product's sustainability, which is another approach to achieving green computing. Indian government should encourage IT companies to make a greater effort to increase product life cycles.

## 7.6 CONCLUSION

Green computing is instrumental for the environment. Organizations can reduce power consumption, as well as carbon dioxide emissions, and positively contribute to environmental stewardship by adopting green computing in their business practices. The key initiatives for green computing emphasize the reusing and recycling processes, power optimization coding algorithm, using virtualization to efficiently use power management and system resources, and cloud computing. The government should create and implement rigorous policies for raising environmental awareness among IT vendors. Green computing's success will depend on developing computing strategies in an innovative way to obtain sustainable products. Since this is an emerging discipline, there is much more research to be carried out, especially within academic sectors.

## REFERENCES

Anandharajan, T.R.V., and M. A. Bhagyaveni. 2011. "Co-operative scheduled energy aware load-balancing technique for an efficient computational cloud." *International Journal of Computer Science Issues* 8(2): 571–576.
Annual Report. 2006. June 2007, Greenpeace International.
Assocham & cKinetics. 2016. E-Waste generation, https://economictimes.indiatimes.com/tech/hardware/indi a-likely-to-generate-52-lakh-mt-of-e-waste-by-2020- study/articleshow /52569725.cms
Automation Business Technologies. 2011. "Green computing through virtualization." http://automationbusinesstech.com/green-computing-virtualization-virginia.
Balaras, C., J. Lelekis, E. Dascalaki, and D. Atsidaftis. 2017. "High performance data and energy efficiency potential in Greece." *Procedia Environmental Sciences* 3: 107–114. doi: 10.1016/j.proenv.2017.03.091.
Boivie. 2008. "TCO takes the initiative in comparative product testing." Accessed October 27, 2020. http://www.boivie.se/index.php?page=2&lang=eng.
Brigden, K., and Santillo, D. 2006. *"Determining the presence of hazardous substances in five brands of laptop computers."* Greenpeace Research Laboratories Technical Note 05/2006; www.greenpeace. to/publications.

Brooks, S., Xuequen Wang, and S. Sakar. 2010. "Unpacking green IT: a review of the existing literature, conference: sustainable IT collaboration around the globe." *16th Americas Conference on Information Systems, AMCIS.* Lima: Peru.

Clancy, H. 2012. "Data center." *GreenTech Pastures.*

Creyts, J., A. Derkach, and S. Nyquists. 2007. "U.S greenhouse gas abetment mapping initiative." *Executive Report.*

Directive 2011/65/EU of the European Parliament and of the Council. 2011. "Restriction of the use of certain hazardous substances in electrical and electronic equipment." *Official Journal of the European Union.*

Directive 2012/65/EU of the European Parliament and of the Council. 2012. "Waste electrical and electronic equipment (WEEE )." *Official Journal of the European Union.*

Dockrill, P. 2016. "Computers will require more energy than the world generates by 2040." *Science Alert.*

Economic Times. 2012. "E-waste management in India." Accessed October 27, 2020. www.economictimes.indiatimes.com/tech/hardware.

Enerdata. 2020. "Global energy trends 2020 edition." https://www.enerdata.net/publications/executive-briefing/expected-world-energy-consumption-increase-from-digitalization.html.

Energuide. n.d. "How much power does a computer use? And how much $CO_2$ does that represent?" Energuide.be. https://www.energuide.be/en/questions-answers/how-much-power-does-a-computer-use-and-how-much-co2-does-that-represent/54/.

European Commission. n.d. "Waste electronic equipment environment." Accessed March 23, 2020. https://ec.europa.eu/environment/waste/weee/index_en.htm.

Fe Bureau. 2018. "Financial express." *Financial Express.* www.financialexpress.com/author/fe-bureau/.

Gao, J. 2014. "Machine learning applications for data center optimization." *Google Report.* https://www.cse.iitk.ac.in/users/cs300/2014/home/~ratnesh/cs300A/techpaper-review/5A.pdf.

Garg, S. K., and R. Buyya. 2012. "Green cloud computing and environmental sustainability." Edited by S. Murugesan and G. R. Gangadharan. *Wiley-IEEE Press Ebook.*

Greig, J. 2018. "Innovation." *TechRepublic.*

*Guardian.* 2017. "Tsunami of data' could consume one fifth of global electricity by 2025." Published December 11, 2017. https://www.theguardian.com/environment/2017/dec/11/tsunami-of-data-could-consume-fifth-global-electricity-by-2025.

Jain, A., M. Mishra, S. K. Peddoju, and N. Jain. 2013. "Energy efficient computing green cloud computing." *International Conference of the Energy Efficient Technologies for Sustainability* 6(6). doi: 10.1109/ICEETS.2013.6533519.

Jaureguialzo, E. 2011. "PUE: the green grid metric for evaluating the energy efficiency in a data center. Measurement method using the power demand." INTELEC, 1–8. doi: 10.1109/INTLEC.2011.6099718.

Jena, R. K. 2013. "Green computing to green business: green technologies and business practices: an IT approach." *IGI Global.*

Jofre, S., and T. Morioka. 2005. "Waste management of electric and electronic equipment: comparative analysis of end-of-life strategies." *Journal of Material Cycles and Waste Management* 7(1): 24–32. doi: 10.1007/s10163-004-0122-1.

Karthikeyan, C. Krishna, and A. Benjamin. 2019. "An exploratory literature review on advancements in applications of cloud and BI; a techno-business leadership perspective." *International Journal of Research in Social Sciences* 4(9): 131–166.

Kaur, M., and P. Singh. 2013. "Energy efficient green cloud: underlying structure." *IEEE International Conference of the Energy Efficient Technologies for Sustainability* 6(6). doi: 10.1109/ICEETS.2013.6533383.

Kaur, T., and I. Chana. 2015. "Energy efficiency techniques in cloud computing –a survey and taxonomy." *ACM Computing Surveys* 48(2).

Kazandjieva, M., B. Heller, O. Gnawali, W. Hofer, P. Lewis, and C. Kozyrakis. 2011. "Software or hardware: the future of green enterprise computing." http://sing.stanford.edu/cs303-sp11/papers/green_computers.pdf.

Kurp, P. 2008. "Green computing – are you ready for a personal energy meter?" *ACM* 51(10): 11–13. doi: 10.1145/1400181.1400186.

Kusnetzky, Dan. 2007. "Virtualization and green computing." Accessed October 27, 2020. http://www.zdnet.com/blog/virtualization/virtualization-and-green-computing/141.

Lundgren, K. 2012. "The global impact of e-waste: addressing the challenge." *International Labour Organization Geneva.*

Marketers Media. n.d. "Global green data center market trend." Accessed March 27, 2019.

Nadaf, R. B., and M. S. Nadaf. 2014. "Green marketing: challenges and strategies for Indian companies in 21st century." *International Journal of Research in Business Management* 2(5).

Nair, K. M., and V. Gopalakrishna. 2009. "Generic web services: a step towards green computing." *International Journal on Computer Science and Engineering* 1(Mar): 248–253.

Nair, K. M., and V. Gopalakrishna. 2009. "CloudCop: putting network-admin on cloud nine towards cloud computing for network monitoring." *IEEE International Conference on Internet Multimedia Services Architecture and Applications (IMSAA).*

Nanath, Krishnadas, and R. R. Pillai. 2017. "The influence of green IS practices on competitive advantage: mediation role of green innovation performance." *Information Systems Management* 34(1): 3–19. doi: 10.1080/10580530.2017.1254436.

Pahlevan, Ali, Maurizio Rossi, Pablo Deivid Valle, and Davide Brunelli. 2017. "Joint computing and electric systems optimization for green data centers." *Handbook of Hardware/Software Codesign*, 1163–1183. doi: 10.1007/978-94-017-7358-4_35-1.

Panda, R. 2013. "E-waste management: a step towards green computing." *International Journal of Environmental Engineering and Management* 4(5): 417–424.

Park, M., and L. Soni. 2019. "India's two-million-tonne e-waste problem has deadly consequences." *Quartz India.*

Prasad, V. K., A. Nair, and S. Tanwar. 2019. "Resource allocation in cloud computing." *BPB Publications, Instant Guide to Cloud Computing* 343–376.

Priya, B., E. S. Pilli, and R. C. Joshi. 2013. "A survey on energy and power consumption models for greener cloud." *IEEE 3rd International Advance Computing Conference* (February): 76–82. doi: 10.1109/IAdCC.2013.6514198.

Ramesh, S., and K. Joseph. 2006. "Electronic waste generation and management in an Indian city." *Journal of Indian Association for Environmental Management* 33(2): 100–105.

Riaz, M., J. Gutierrez, and M. Pedersen. 2009. "Applied sciences in biomedical and communication technologies." *2nd International Symposium, Isabel.*

Riyaz, A, Sheikh, and U. A. Lanjewar. 2010. "Green computing- embrace a secure future." *International Journal of Computer Applications* 10(4): 22–26. doi: 10.5120/1469-1984.

Schipper, I., and E. de Haan. 2005. "ICT hardware manufacturing sector SOMO ICT Sector." *CRC Report.*

Sego, L. H., A. Marquez, A. Rawson, T. Cader, K. Fox, W. Gustafson, and C. J. Mundy. 2012. "Implementing the data center energy productivity metric." *ACM Journal on Emerging Technologies in Computing System* 8(4): 30. doi: 10.1145/2367736.2367741.

Sheth, H. 2019. "E-waste management may end up as just junk sans policy support: Experts." *Business Line.*

Sinha, R., N. Purohit, and H. Diwanji. 2011. "Power aware live migration for data centers in cloud using dynamic threshold." *International Journal of Computer Technology Application* 2(6): 2041–2046.

Soomro, T. R., and M. Sarwar. 2012. "Green computing: from current to future trends." *World Academy of Science, Engineering and Technology* (6): 3–20.

Statista. "$CO_2$ emission in the year 2020." Statista Research Department.

Taufiq-Hail, G. A., H. Ibrahim, and S. Affendi. 2017. "SAAS cloud computing as a means of green it acceptance model: a theory of planned behavior model at Malaysian Public Universities' context." *Journal of Information System and Technology Management* 2(July): pp. 1–17.

Tiwari, Shobhit. 2011. "Need of green computing measures for Indian IT industry." *Journal of Energy Technologies* 1(4), 18–25.

Wang, L., and S. U. Khan. 2011. "Review of performance metrics for green data centers: a taxonomy study." *Journal of Supercomputing* 63(3): 1–18.

Weissberger, Alan. 2019. "AT&T announces cloud partnership with Microsoft 1 day after similar deal with IBM." *IEEE Communications Society Technology Blog.*

Wheeland, M. 2007 "Green computing at Google." *GreenTech Pastures.*

Wikipedia. N.d. "United States Environmental Protection Agency." Accessed October 20, 2020. https://en.wikipedia.org/wiki/United_States_Environmental_Protection_ Agency.Access.

Wipro. 2010. "Wipro at the forefront of green IT revolution: introduces 100% recyclable toxin free greenware PCs." Wipro.com. Accessed October 20, 2020. https://www.wipro.com/en-IN/newsroom/press-releases/archives/wipro-at-the-forefront-of-green-it-revolution–introduces-.

Worthington T. 2017. "ICT sustainability: assessment and strategies for a low carbon future." ISBN 9781326967949.

Yamini, R. 2012. "Power management in cloud computing using green algorithms." *IEEE-International Conference on Advances in Engineering, Science and Management* (March): 30–31.

Yuventi, J., and R. Mehdizadeh. 2013. "A critical analysis of power usage effectiveness and its use in communicating data center energy consumption, energy and buildings." *Elsevier* 64: 90–94.

# 8 Technological Innovations for Environmental Sustainability

*Tapalina Bhattasali and Xavier Savarimuthu, SJ*

## CONTENTS

## 8.1 INTRODUCTION

Over the past few decades, environmental degradation has increased at an alarming rate. The Earth has been adversely affected by issues like global climate change, air pollution, waste disposal, water pollution, and many others. Irresponsibility of people has led to the exacerbation of hazards like floods, cyclones, forest-fires, droughts, and even pandemics. There is an urgent need to find optimal solutions to protect the environment. Although inappropriate use of technology can negatively affect the environment, recent tools and technologies play a key role in sustainable development. To create a positive correlation with environmental sustainability, technology needs to be used deliberately in order to avoid damaging natural resources. Recently, technological innovations have seen tremendous advances, especially in the areas of artificial intelligence (AI), machine learning, Internet of Things (IoT), sensors, robotics, computer-vision, cameras, drones, and blockchain. These smart technologies have opened new doors for effectively responding to environmental hazards and even try to prevent these hazards from occurring. The objective is to explore the concept of green technology that can provide eco-friendly solutions, protect the environment, conserve natural resources, and repair the damage. This chapter discusses the use of innovative technologies in various sectors to achieve environmental sustainability.

## 8.2 RECENT TRENDS AND TECHNOLOGIES FOR A SUSTAINABLE ENVIRONMENT

Recent trends and technologies can make a positive difference when seeking innovative solutions in the fight against environmental hazards. Consequently, the use of IoT, sensors, drones (useful due to being lightweight, simple, low-cost, and not requiring fuel), green robots, AI, machine learning, machine vision techniques, computer aided design (CAD), and advanced digital technologies is becoming popular in various application areas to reduce human intervention and produce eco-friendly solutions.

### 8.2.1 AGRICULTURE AND FARMING

Technology has the potential to bring about the use of agricultural methods that are less harmful to the environment (Cropin 2017). AI can analyze the temperature, weather and soil conditions, or water use to make better real-time decisions (Walch 2019). AI and IoT can be used to monitor crops and soil to maximize crop production without affecting the environment (Talaviya et al. 2020). Smart monitoring devices, along with sensors, can monitor crop growth and collect relevant data on hydration, disease, and plant nutrition. Then, these data can be used to determine irrigation patterns to recommend the best watering cycles. Any anomalies can be immediately detected and resolved. Using innovative technology, farmers can forecast the weather, identify crop disease, analyze crop inputs, plan for optimized resource use by determining their choice of crops with high nutritional content, and farmers can decide on the choice of seeds to generate a higher yield reducing the use of harmful pesticides. Global positioning systems (GPS) and short message services (SMS) can be used to protect farms from damage, e.g. authorities are alerted when any animals, such as wild elephants, are too close to the farm. Geographical information system (GIS) is generally used for precision farming. The practice of digitally mapping agricultural land can be combined with various data related to topography, contour, and other statistical information to analyze soil. GIS, along with historical data and sampling, can also be used to make decisions about what and where to plant. There are various smartphone apps, like the Indian government created app Meghdoot ("Cloud messenger"), which forecast rainfall, temperature, wind speed and direction, and humidity. Robots that are used for crop harvesting use sensors and computer vision to determine the optimal time to collect fruits and vegetables. Additionally, robots are used to determine the optimal level of synthetic or natural soil fertilizers for maximum productivity. Robots can be used to automate tasks, such as planting, sowing, and watering within a short-time window and with high precision. Through wireless technology, IoT, and AI, automated and no-contact systems can monitor plant growth and health status in any remote area. This system can avoid the untimely deaths of young planted saplings. Drones, along with android-based devices with electronic circuitry, remote servers, and cloud databases, can be used to track, locate, and monitor the planted saplings' health. To monitor the plant's growth and health, the system can continuously analyze basic parameters, including temperature, soil's moisture level, humidity, concentration of $CO_2$, and the plant's leaf color.

There are also robots that are designed to spray pesticides in agricultural fields. Fully deployed robots can move around in a farm, take photographs of plants, and diagnose any disease. If the robot detects a disease in a plant, it will select an appropriate pesticide or medicine from its tank and spray it on that plant. Robots, like the TreeRover (DESIGN INDABA 2015), can plant trees. Drones spray plants in a way that reduces the amount of fertilizer used, as well as protect the environment from pollution (Sylvester 2018). Drones are becoming an invaluable tool to monitor water levels, crop, and livestock. Detailed information from high-resolution images can be used to analyze crop health, improving the crop yield. Sophisticated UAV (unmanned aerial vehicle) can be used to create three dimensional images of the landscape in planning for the future. Drones are useful, faster, and cheaper in fighting deforestation compared to planting trees by hand.

In the fight against climate change, drip irrigation protects the environment from inefficient crop irrigation methods. Drip irrigation provides a sustainable way to reduce waste and use of agricultural water while increasing crop yield and profitability. This irrigation method delivers an exact amount of water to the plant instead of incurring water losses by spreading the water around the surrounding area. To provide a fully automated system, drip irrigation systems can easily be combined with an environment monitoring system through weather stations and soil sensors. Thus, adopting new strategies for the effective management of farm water and encouraging drip irrigation improve agricultural water sustainability.

In livestock farming, robots can be used for managing a herd of domestic animals or milking cows (Duckett et al. 2018). An automated milking system milks dairy cattle without human intervention. The automated milking procedure is controlled by an agricultural robot, herd management software, and a specialized computer. Data analytics is used to improve herd management, as well as to obtain optimal milk yields.

## 8.2.2 WASTE MANAGEMENT

Recent tools and technologies improve waste management procedures, from collecting garbage to transporting, sorting, and recycling in a cost-effective way. Smart bins with IoT sensors can measure the bin's waste level and send the data to the central waste disposal system to categorize the type of garbage and determine appropriate disposal methods. The sensors in smart bins ensure that the trash is cleared before it overflows, and trash pickup is only possible when the trash cans are filled. AI, along with analytics, can integrate various types of collected waste with fleet management systems. Smartphone apps can provide citizens with the location of the nearest empty trash bin to avoid the problem of overflowing bins. These methods can help municipalities make data-driven decisions to optimize waste collection frequencies, routes, and times (Pardini et al. 2020).

Robots make recycling easier by identifying and separating various types of waste and transferring the waste into the garbage bin (Apoorva et al. 2017). Additionally, robots are designed as a hospital-waste management solution to separate, transport, and treat different types of waste from the hospital. Robots can perform various tasks like vacuuming, cleaning, mowing, and sorting garbage. AI-enabled garbage collection robots can use GPS to collect household waste. This type of robot collects data from its attached sensors and uses AI algorithms to determine the optimal garbage collection route and process. Waste management robots include waste treatment and recycling and the automated processes involved in waste management and waste treatment can reduce $CO_2$ emissions. Robots quickly and efficiently sort waste materials without incurring power and cost overhead (Pardini et al. 2020).

## 8.2.3 WATER MANAGEMENT

AI can help reduce water pollution. Autonomous floating garbage trucks powered by AI take apart the Great Pacific Garbage Patch. Machine learning can enable a quick and effective response to marine litter (Recode 2018). Machine learning algorithms can also easily monitor, analyze, and predict the conditions that negatively impact the oceans due to industrial activities, illegal fishing, coral bleaching, and disease outbreaks (Recode 2018). Using AI has improved water and waste management systems. AI can manage use, monitor water quality, and predict maintenance needs. IoT, machine learning, and blockchain have the potential to create decentralized water systems that help keep drinking water clean and plentiful (Recode 2018). AI and robotics can be used to clean and restore rivers. Using an automated water vehicle to clean rivers can eliminate chemical effluents and other types of trash and waste that float on the river's surface (Recode 2018).

IoT- and AI-based autonomous garbage collection trucks remove trash from the oceans. Machine learning algorithms analyze conditions that affect the ocean, such as solid waste disposal and discharged industrial effluents. Technology can help to improve water quality by instantaneously monitoring pollution levels (Recode 2018).

Robots can ingest microbes in the ocean and convert them into energy. This type of robot can remove chemicals from rivers and oceans and reduce the impact of oil leaks from tankers and ships. Robots are used for collecting waste from bodies of water in a cost-efficient and time-efficient way (Recode 2018).

### 8.2.3.1 Arsenic Detection in Water

A prediction model with AI can detect arsenic pollution in drinking water (Nature India 2020). AI plays an important role in identifying sources of drinking water for areas that suffer from groundwater pollution problems. The objective is to provide locations of safe groundwater, which is the primary source of drinking water in many regions.

Artificial neural networks (ANN) can be used to estimate the extent of arsenic contamination in the groundwater (Purkait et al. 2008). The input data are collected from areas impacted by arsenic. Each data sample contains the amount of arsenic contamination found in the groundwater and relevant geochemical parameters, such as total dissolved solids (TDS), observed values for pH, salinity, specific conductivity, dissolved oxygen (DO), redox potential (Eh), and depth of tube well water. The expected output for the ANN training phase is the amount of arsenic contamination observed in the groundwater. The ANN model has the capacity to predict better results for estimating arsenic contamination in groundwater.

### 8.2.3.2 Smart Water Management

Wireless IoT is employed in smart water management solutions to facilitate the movement of water along designated routes. In this data-driven setup, high-end sensors placed across reservoirs and overhead tanks are employed to establish the water level in the tanks or reservoirs (Müller-Czygan 2020). IoT is employed in smart water management solutions to ensure water reaches the target destination and to conserve water resources. Through tracking real-time water flow, sensor-based IoT devices optimize water use and reduce the demand for water. Users can directly receive information about their water consumption pattern and data on their mobile phones. The software takes into account issues like leakages, abnormal use, and open taps to help people make immediate decisions and decrease their water consumption (Müller-Czygan 2020).

### 8.2.4 ENERGY MANAGEMENT

The use of domestic energy can be reduced through smart home energy management systems that help people find ways to save power. For example, lowering the thermostat by 1°C can reduce bills by about 10%. Real-time energy monitoring apps connect to smart meters that have the ability to remotely switch devices off, which helps customers understand and better manage their energy use. Smart technologies automatically manage energy use for customers (Verma 2019).

Robots can simplify the processes involved in generating renewable energy. They can automate the task of cleaning solar panels to provide maximum output and generate high amounts of energy. Robots can also generate their own electricity while carrying out assigned tasks (Verma 2019).

Renewable energy technology, such as solar panels and wind turbines, is becoming more efficient and cost effective. AI can help increase the ubiquity of renewable energy and decrease dependency on polluting fossil fuels. AI-enabled smart grids can improve the way people use electricity in their offices or homes. Smart meters save energy and provide a better understanding of use to energy providers. Energy providers can make immediate adjustments to enhance efficiency. Through smart grids, the energy-delivering process can be improved and reduce the amount of waste (Verma 2019).

### 8.2.5 AIR POLLUTION DETECTION

AI can quickly and accurately monitor pollution levels and identify the sources of air quality issues. For example, targeted remediation can be possible during gas leaks by using self-organizing mesh network technology and smart sensors with machine learning. Additionally, autonomous vehicles can reduce oil consumption and related greenhouse gas emissions. Autonomous ride shares and dynamic bus routing can control the number of cars on roads. AI-enabled traffic lights can adjust the flow of traffic to minimize driving time (Recode 2018).

### 8.2.6 ENVIRONMENT MONITORING AND LAND MANAGEMENT

Drones are very effective for monitoring the environment, and they make it much easier to reach environmentally critical places. Through aerial mapping, forestry regions can be conserved and protected from illegal logging. Drones effectively monitor the populations of different species, determine their ranges, and stop poachers. Precision mapping is possible due to drones taking close-up footage. Drones can be sent to various landscapes, from forests to glaciers to ocean waters, with sensors and thermal imaging equipment (Recode 2018).

Effective land management can be achieved by using drones to collect data, including information on weather and soil moisture. This process assists optimal planning and monitors the future growth and health of crops, providing a clear picture to the landowners.

### 8.2.7 WILDFIRE DETECTION

IoT, AI, and drones are capable of detecting the early signs of emerging wildfires. This technology makes it possible to prevent or minimize the scale of a disaster. IoT sensors can monitor the parameters for detecting wildfires, which include temperature, $CO_2$, CO, and humidity, and then use this information to prevent wildfires. To plan for wildfire prevention, vital data need to be collected from multiple sources, such as IoT sensors, cameras, satellite images, weather data, archived data, social media, and modeling tools. A diverse range of data is gathered and analyzed using AI to predict the likelihood of a wildfire or even prevent a wildfire. AI-based autonomous drones can quickly and accurately measure fire. Drones are also used to survey and extinguish fire, and they can be used to restore infrastructure and save lives (Foverskov 2020). Robots help avoid risking the lives of firefighters by quickly controlling wildfires. Robots can be remotely controlled to efficiently detect and extinguish fires through GPS, computer vision, heat sensors, and AI technologies. These components are very useful in areas that are unsafe for humans. With their robotic arms and camera sensors, robots can save lives and rescue people caught in wildfires (Foverskov 2020).

## 8.3  CASE STUDIES

### 8.3.1  SMART WATER SOLUTIONS

AI systems can help smart cities to solve the underground water crisis (Smartcity 2019). The SmartCover System in California, US, has developed robust technological solutions. This company combines sensors, satellite communication, analytics, and event notification platforms to immediately report sewer conditions. AI pattern recognition software manages issues related to sewer infrastructure, eliminating the need for a person to enter a manhole. Strategically positioned remote field units have been developed to identify blockages before an overflow occurs. These field units can also detect storm water infiltration and prevent sewer spills before they back up into homes and businesses.

The AI CENTAUR System (Cost Effective Neural Technique to Alleviate Urban Flood Risk) CORDIS, 2020 can manage water flow in cities. The system controls the water flow from one end of the city to the other by being integrated with sensors that monitor water level at gates that can be remotely operated to control the flow during extreme weather conditions, preventing flooding in specific areas and water from spilling on the streets. It works without human involvement and learns from its mistakes.

Fluid Robotics (http://www.fluidrobotics.com/) is an AI-powered solution to diagnose problems with water and wastewater pipeline networks to address water loss and water pollution. A multi-sensor robotic system is deployed for better managing the sewer network and reducing pollution in bodies of water. Fluid Robotics uses AI capabilities to perform fault detection without human intervention, and it supports seamless data acquisition and visualization. Fluid Robotics works as a robot-as-a-service business model that includes the additional feature of topographic surveys, flow surveys, hydraulic modeling, and hydrological modeling.

Normally, about 600 mL of water is used in a single hand wash. During the current COVID-19 pandemic as of the writing of this manuscript, the requirement of frequent hand washing may give rise to a water crisis in the near future. In India, nozzles have been developed to cut water waste by 95% (Wangchuk 2019). The nozzles change the tap water into a fine, gentle, and mist-like spray requiring only 15–20 mL of water per hand wash (Wangchuk, 2019). These nozzles, along with sensors, can be installed in washbasin taps. When the sensors detect hands, the water flow is reduced to a mist, which brings down water consumption by 80% and prevents water waste.

**TABLE 8.1**

**Comparative Analysis between Drip Irrigation and Conventional Irrigation**

|  | Crop Yield (kg/acre) | | | Water Use (kg/acre/cm) | | |
|---|---|---|---|---|---|---|
|  | Surface | Drip | % More | Surface | Drip | % Savings |
| Sugarcane | 30,000 | 70,000 | 133.3 | 9800 | 4960 | 49.3 |
| Cotton | 904 | 1700 | 88.0 | 3600 | 1680 | 46.6 |
| Onion | 14,625 | 22,500 | 53.8 | 2080 | 1120 | 46.1 |
| Potato | 6060 | 10,880 | 79.5 | 2400 | 1100 | 54.1 |
| Chili | 912 | 1520 | 66.6 | 1708 | 980 | 42.6 |

*Source:* Netafim India.

## 8.3.2 DRIP IRRIGATION

Drip irrigation is the most efficient and environment friendly technological innovation that addresses many critical sustainability challenges. Advanced drip irrigation systems can enhance productivity and output, while using less water and resources (Rosenbaum and Markovitz 2012). Table 8.1 shows a comparison between drip irrigation and surface irrigation for crop yield and water use.

According to Figure 8.1(a), crop yields are far better in drip irrigation compared to conventional irrigation. Figure 8.1(b) shows that water usage is lesser in case of drip irrigation, which results into savings of water.

## 8.3.3 SANITARY WASTE MANAGEMENT

In India, disposing sanitary napkins is a daunting challenge. Of discarded pads, 98% end up in bodies of water and landfills, and pads take a long time to decompose. If incinerated, a pad can release 380 kg carbon into the atmosphere. PadCare Lab, based in Pune, India, provides a safer, greener, and more efficient method of sanitary waste disposal that mitigates harmful environmental impacts. Climate-friendly technologies are used to manage sanitary waste. Through SANECO, PadCare Lab converts hazardous sanitary napkin waste into eco-friendly waste (WIPO I GREEN, 2020). UVECO, a UV-based and touchless sanitary pad collection bin, is generally installed in washrooms and automatically opens its lid when someone moves their hand over the sensor. The collected sanitary napkins are then transferred to the centralized processing unit, SANECO, which disinfects and separates the waste. The sanitary pads' residual by-products can be recycled into paper or plastic pellets (Kleshchenko 2020).

## 8.3.4 GO GREEN WITH IoT APPLICATION AREAS

An IoT-enabled company Artveoli, focuses on fresh air and efficient air ventilation. Their air panel device uses photosynthetic cells to convert $CO_2$ into $O_2$. Another

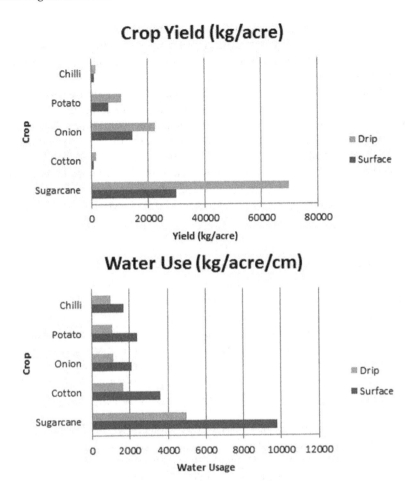

**FIGURE 8.1** (a) Comparison between drip and surface in the context of crop yield. (b) Comparison between drip and surface in the context of water use.

IoT-enabled company, HydroPoint (HydroPoint 2020) uses sensor data analytics to provide smart water management for irrigation applications. WeatherTRAK (HydroPoint 2020), a cloud-based irrigation controller, operates in real-time. Embue (Embue 2020) uses IoT technology to analyze collected data for immediate energy savings and predictive monitoring. With Embue's remote smartphone app, residents in smart buildings can control their home equipment, including thermostats, sensors, and leak detectors. CityMapper is an IoT-enabled car-sharing service that can reduce traffic and is less harmful to the environment (Thomas 2020).

### 8.3.5 GO GREEN WITH DRONE APPLICATION AREAS

Drones are useful for mapping purposes by providing clear images of land, such as conserved wetlands or tropical rainforests (Edie 2016). Small drones can be used for maintaining sources of renewable energy. At a minimal monetary and environmental cost, these drones can send three-dimensional images of turbine blades, real-time videos of power cables, and HD live video of hydro-electric dam walls to inspect for damage (Edie 2016).

Unmanned vehicles are becoming increasingly used to combat environmental disasters. Information from the affected area is necessary to begin the process of disaster relief. Drones can analyze the situation and recommend how people should use resources to mitigate damage and save lives (Edie 2016). Farming drones can be used as an agricultural sustainability solution. A research team at Timiryazev State Agrarian University in Moscow designed a drone to capture high-resolution images for a wheat fertilization project. These images can create a custom map to optimize their nitrogen application, leading to a 20% reduction in applied nitrogen (Edie 2016).

### 8.3.5.1 Environmental Impact of Drones, Electric Motorcycles, and Gasoline Motorcycles

The impact of drones, electric motorcycles, and gasoline motorcycles on the environment has been analyzed and compared regarding mobility, applicability, and efficiency of delivery. There is a growing interest in electric motorcycles because they have low emissions. However, drones are the greener solution for delivery. Drones generally deliver lightweight products in specific zones where normal delivery may not be possible. Since electric motorcycles have a low fuel cost and low carbon emissions, they can replace gasoline motorcycles. Electric motorcycles can also carry a greater quantity of products than drones (Park et al. 2018).

The environmental impact of drones, electric motorcycles, and gasoline motorcycles was calculated using global warming potential (GWP) and particulate releases when delivering goods within a distance of 1 km. On a scale of highest to lowest for GWP, these three delivery methods are ranked in the order of gasoline motorcycle, electric motorcycle, and drone. In ranking the particulate releases from highest to lowest, the order is electric motorcycle, gasoline motorcycle, and drone. Electric vehicles are eco-friendly based on greenhouse gas emissions, but electric motorcycles have a worse environmental impact than gasoline motorcycles in terms of particulates (Park et al. 2018).

### 8.3.6 Greener Online Shopping

Ordering products online is a greener shopping method as compared to driving from home to the store and back. A delivery van is packed with items to be delivered to different houses and uses less fuel than having a number of cars driving between home and the store. Delivery vans that bring items to the customer's doorstep emit 20–75% less $CO_2$ per customer on average than the scenarios where customers drive their cars between their home and the store. The result can vary based on the scenarios of online delivery, such as whether the customer selects the delivery slot or the online store sets the delivery time based on the proximity of other online orders. If the store can decide the delivery time, then the van can follow a more efficient route, with carbon savings between 80% and 90%. If customers select the delivery slot, then the van spends comparatively more time on the road, still resulting in carbon savings between 17% and 75%. According to Figure 8.2(a) and (b), the delivery van is the greener solution compared to a customer driving their personal car between their home and shopping location in both the scenarios (Plumer 2013). Figure 8.3 shows that $CO_2$ emissions of delivery vans per customer is significantly less if the delivery route is selected by proximity assignment instead of random selection of customers.

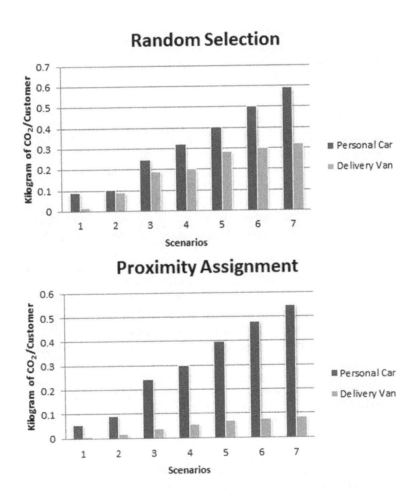

FIGURE 8.2    (a) $CO_2$ emissions between personal car and delivery van based on random selection. (b) $CO_2$ emissions between personal car and delivery van based on proximity assignment. (Source: Economic Policy.)

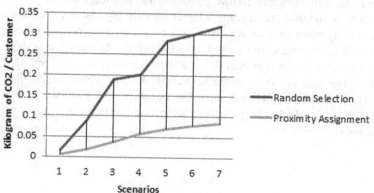

**FIGURE 8.3** Comparative analysis of $CO_2$ emissions of delivery vans based on random selection and proximity assignment.

## 8.4 CONCLUSIONS

The availability of advanced technologies helps simplify problems related to environmental hazards and slow down the negative environmental impacts. Technological innovation can provide faster, efficient, and adaptable solutions. AI has the potential to significantly accelerate sustainability efforts (Recode 2018). In every sector, AI is emerging as the key that anyone can use to protect the natural environment. AI can detect plant diseases, predict poacher routes, monitor erosion, identify species, and track animal migration. To prevent damaging natural resources, AI, machine learning, IoT, drones, and robotics can be used to achieve better monitoring and understanding. Robots with additional advanced technology can strategically help the environment in sustainable development. The increasing demand for green robots is because they carry out environment-friendly tasks with the utmost efficiency without tiring. Drones fill the gap between satellites and ground surveying. They enable fast access to remote areas to analyze situations and obtain data. Drones are also used to provide food and medical supplies to people in need and to plant trees in the fight against deforestation.

The smart technologies discussed in this chapter can be used to predict and prevent natural catastrophes. Society needs to be creative and innovative in harnessing the use of technology for environmental sustainability and to restore ecosystem potential.

## REFERENCES

Apoorva S., Chaithanya, Rukuma S. Prabhu, Saiswaroop B. Shetty, Denita D'Souza. 2017. Autonomous Garbage Collector Robot. *International Journal of Internet of Things* 6(2). Available Online At: http://article.sapub.org/10.5923.j.ijit.20170602.06.html.

CORDISEU Research Results. European Commission. Horizon 2020. 2020. "Cost Effective Neural Technique for Alleviation of Urban Flood Risk" Available Online At: https://cordis.europa.eu/project/id/641931/results.

Cropin. Technology Solution. 2017. "An Introduction to Smart-Farming." Available Online At: https://www.cropin.com/smart-farming/#:~:text=Smart%20farming%20is%20 a%20big,processed%20to%20produce%20farm%20insights.

DESIGN INDABA. 2015. "Autonomous Tree-Planting Robot to Tackle Forest Restoration" https://www.designindaba.com/articles/creative-work/autonomous-tree-planting- robot-tackle-forest-restoration.

Duckett, T., S. Pearson, S. Blackmore, and B. Grieve. 2018. "Agricultural Robotics: The Future of Robotic Agriculture." *UK-RAS White Paper.* Available Online At: https:// arxiv.org/ftp/arxiv/papers/1806/1806.06762.pdf.

Edie. 2016. "Five Ways Drones are Being Used to Help the Environment." Available Online At: https://www.edie.net/news/8/Drone-technology-environmental-sustainability-impact- for-the-UK/.

Embue. 2020. "Apartment Building Intelligence. Available Online At: https://embue.com/.

Foverskov, L. 2020. "Preventing Forest Fires with Smart Technology." Available Online At: https://orangematter.solarwinds.com/2020/04/13/preventing-forest-fires-with-smart- technology/#:~:text=Over%20the%20last%20few%20years,and%20effectively%20 respond%20to%20bushfires.

HydroPoint. 2020. "360 Degree Smart Water Management Technology". Available Online At: https://www.hydropoint.com/.

Kleshchenko, L. 2020. "Women in Green: Tackling Hygiene Waste through Innovation." WIPO| GREEN. Available Online At: https://www3.wipo.int/wipogreen/en/news/2020/ news_0007.html.

Müller-Czygan, G. 2020. "Smart Water—How to Master the Future Challenges of Water Management." Available Online At: https://www.intechopen.com/online-first/ smart-water-how-to-master-the-future-challenges-of-water-management.

Netafim. 2018. "Drip Irrigation." Available Online At: https://www.netafim.com/en/ drip-irrigation/

Nature India. 2020. "AI to Detect Risk of Groundwater Arsenic Exposure." Available Online At: https://www.natureasia.com/en/nindia/article/10.1038/nindia.2020.147.

Pardini, K., J. J. P. C. Rodrigues, O. Diallo, A. K. Das, V. H. C. D. Albuquerque, and S. A. Kozlov. 2020. "A Smart Waste Management Solution Geared towards Citizens." *Sensors* 20(8), 2380. Available Online At: https://www.mdpi.com/journal/sensor.

Park, J., S. Kim, and K. Suh. 2018. "A Comparative Analysis of the Environmental Benefits of Drone-Based Delivery Services in Urban and Rural Areas." *Sustainability* 10 (3): 888.

Plumer, B. 2013. "Economic Policy. Study: Ordering Groceries Online is Greener Than Driving to the Store." Available Online At: https://www.washingtonpost.com/news/ wonk/wp/2013/04/30/ordering-groceries-online-is-greener-than-driving-to-the-store/.

Purkait, B., S. Kadam, and S. K. Das. 2008. "Application of Artificial Neural Network Model to Study Arsenic Contamination in Groundwater of Malda District, Eastern India." *Journal of Environmental Informatics.* 12 (2): 140–149.

Recode. 2018. "How AI Can Help Us Clean Up Our Land, Air, and Water". Available Online At: https://www.recode.net/ad/18027288/ai-sustainability-environment.

Rosenbaum, A., and E. Markovitz. 2012. "Drip Irrigation – A Key Pillar in Sustainable Productivity." Available Online At: https://events.development.asia/system/files/ materials/2012/04/201204-drip-irrigation-key-pillar-sustainable-productivity.pdf.

Smartcity. 2019. "3 Ways AI Is Helping Solve Water Crisis Beneath the Smart Cities." https:// www.smartcity.press/water-crisis-solutions-with-ai/.

SmartCover. 2020. Available Online At: https://smartcoversystems.com/.

Sylvester, G. 2018. "E-Agriculture in Action: Drones for Agriculture. Food and Agriculture Organization of the United Nations and International Telecommunication Union." http://www.fao.org/3/I8494EN/i8494en.pdf.

Talaviya, T., D. Shah, N. Patel, H. Yagnik, and M. Shah. 2020. "Implementation of Artificial Intelligence in Agriculture for Optimisation of Irrigation and Application of Pesticides and Herbicides." *Artificial Intelligence in Agriculture* (4): 58–73. Available Online At: http://www.sciencedirect.com/science/article/pii/S258972172030012X.

Thomas, M. 2020. "13 Examples of IoT In Environmental Sustainability You Should Know." Available Online At: https://builtin.com/internet-things/iot-environment-sustainability-green-examples.

Verma, S. 2019. "Smart Energy Management: First Step Towards IoT Adaptation." *Energycentral.* Available Online At: https://energycentral.com/c/iu/smart-energy-management-first-step-towards-iot-adaptation.

Walch, K. 2019. "How AI Is Transforming Agriculture." *Cognitive World.* Available Online At: https://www.forbes.com/sites/cognitiveworld/2019/07/05/how-ai-is-transforming-agriculture/#a86f20c4ad10.

Wangchuk, R. N. 2019. "Chennai Engineers Develop Nozzles to Cut Water Wastage By 95%, Save 35 Litres/Day." *The Better India.* Available Online At: https://www.thebetterindia.com/188892/chennai-water-crisis-engineer-invent-tap-nozzle-saves-litres-home-india/.

WIPO|GREEN. 2020. "Climate-Friendly Management of Sanitary Waste", Available Online At: https://www3.wipo.int/wipogreen/en/news/2020/news_0015.html.

# Section III

## Environmental Sustainability

# 9 Green Transportation
## *Exploring Carpooling as an Environmentally Sustainable Solution*

*Romit S. Beed, Suhana Biswas,*
*Sunita Sarkar, and Arindam Roy*

## CONTENTS

## 9.1 INTRODUCTION

As countries use technology in their quest for industrial growth and societal development, expanding road networks and digital networks is a necessity. Digital growth has been achieved through developing high-speed connectivity networks and devices that have been made available among the masses. Physical point-to-point connectivity is only possible through developing road networks' infrastructures. However, the growth in road networks is expensive, time consuming, and suffers from legal complications associated with land acquisition. This has prompted an exponential increase in vehicular traffic on the existing road space, leading to the concept of vehicle sharing as a green, environmentally sustainable solution.

Research has clearly revealed that the environment and society have been badly affected by poor planning for road transportation and implementation, causing irreparable damage to the planet. Transportation and environment can be oppositional in real life – transportation provides significant socioeconomic upliftment at the cost of environmental degradation. Though transportation development is necessary to meet growing mobility demands, it results in increased environmental externalities and is a major cause of pollution. Key strategies to achieve a sustainable transport system include vehicle route planning and traffic regulation.

Sustainable development is the process of meeting a nation's socioeconomic development goals through properly sustaining and conserving natural systems, natural resources, and ecological systems. It is preferred that society continue to use natural resources without natural systems becoming rapidly deteriorated or destabilized, while making provisions for future generations to continue using these resources in a comparable manner to the present populace. Thus, sustainable development is a multidisciplinary activity involving complex interconnections between the environment, economy, society, and welfare of future generations.

Current developments in vehicle sharing techniques, the greater use of public vehicles, and non-motorized transportation techniques promote sustainable transportation. However, the techniques used for transportation sustainability have restrictions imposed on them by the existing infrastructure. In order to achieve sustainability, positive steps that need to be taken include implementing a demand-based transport system, improving the transport supply system, expanding the current infrastructure to meet the rising demand, and improving the communication system. A higher level of sustainability can be achieved through greater flexibility, lower costs, improved existing public transit facilities, service coverage, and an increase in frequency during demand periods.

Expanding cities and industries put immense stress on the city transport systems. The uncontrolled growth of vehicles on city roads result in high levels of noise and air pollution in urban areas. As indicated by the Canadian national environmental agency, Environment and Climate Change Canada, air contamination from vehicular discharges leads to serious medical conditions, like cardiovascular and respiratory ailments, hypersensitivities, and neurological impacts (Government of Canada n.d.). To achieve environmental sustainability, increasing the shared use of vehicles is an effective tool, particularly in under developed and developing nations that are still in the early stages of exploring these options (Mulders 2012; Manzini and Pareschi 2012). To establish effective transportation frameworks in cities, it is essential to implement various modes of travel that positively affect inhabitants' mobility

(Parezanović et al. 2019). One of the most effective methods is carpooling because it facilitates the greater use of a vehicle and results in a drastic reduction in pollution levels, as well as providing an effective solution to parking problems.

As per Cambridge Dictionary, carpool refers to "a group of people who travel together, especially to work or school, usually in a different member's car each day, usually taking turns to drive" (Cambridge Dictionary n.d.). Collins English Dictionary adds that carpooling is "an arrangement that avoids the unnecessary use of several under-occupied vehicles" (Collins English Dictionary n.d.). Carpooling, also referred to as ride sharing, was initially started in the 1970s due to high prices and lesser availability of fuel in the United States. People shared their rides with colleagues, acquaintances, neighbors, and friends. It assumes that when more than one person shares the journey, the cost associated with driving is significantly reduced. Based on this, there are two principal elements of carpooling: (1) a group of individuals sign up to share the expenses of a specific journey in a shared car, and (2) the owner of the car agrees to offer his own vehicle to be shared by multiple passengers. It is clear from Figure 9.1(a) that the primary reason for choosing carpooling is the reduced

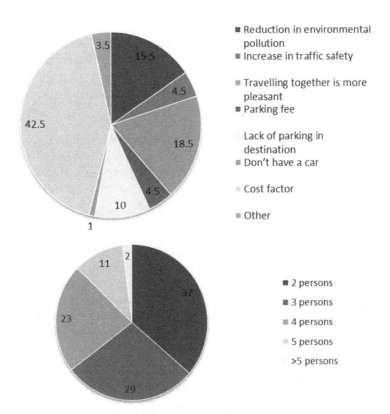

**FIGURE 9.1** (a) Factors favoring popularity of carpooling; (b) Car-wise count of occupiers among carpoolers (Rijavec et al. 2020).

expenses. Apart from the reduced costs, the other two reasons for why people opt for ride sharing are the pleasure of traveling together and the reduction in pollution. While traveling together, two people travel together in 37% of the cases, followed by three people traveling together in 29% of cases, as per Figure 9.1(b). The occupancy cost tends to gradually decrease as the number of persons increases.

There are two primary approaches to carpooling. The first approach is a static carpooling method, wherein the passengers provide their ride-sharing requests hours ahead of time for a future transportation need, and shared rides are scheduled before the journey begins. The second approach is a newer and more advanced concept with a dynamic approach that is superior to the static approach. In this approach, the service provider uses a computerized matching algorithm in real time to coordinate passengers to the appropriate vehicles or cars, depending on various parameters like pick-up or drop-off locations, seat availability in a vehicle, route of the trip, etc.

Additional common carpooling practices will be discussed in the following sections.

### 9.1.1  INFORMAL CARPOOLING

Strangers and neighbors travel together when their paths of travel are along the same routes. Here, the travelers pay a titular fare to compensate for the service provider's costs. In many cases, the drivers pick up the passengers at different points along the travel route.

### 9.1.2  VANPOOLS

Vanpools are a component of the transport framework that permit groups of individuals to share their rides like in a carpool however, a vanpool achieves a larger scope with simultaneous savings in fuel and vehicle working expenses. Vehicles might be provided by an individual traveler or by a group of people in collaboration with different government and private programs, for instance.

### 9.1.3  AD HOC CARPOOLING

The travelers request rides as needed, and usually an automated system matches the riders to the best-match vehicles, depending on pick-up and drop-off locations, service price, etc. Due to the large-scale information and communication technology (ICT) tools available, this form of dynamic carpooling has proved to be immensely beneficial and easy to access for both the customers and the service providers.

## 9.2  ENVIRONMENTAL BENEFITS OF CARPOOLING

The fundamental advantage of carpooling lies in improving the quality of life in urban cities through the resulting decrease of carbon emissions. Carpooling also guarantees a decrease in the number of vehicles in use and provides the most efficient use of the vehicle's available seating, resulting in lower fuel consumption. According to the U.S. Environmental Protection Agency, almost 25% of the country's total greenhouse gas pollution stems from the use of light-duty vehicles (United States Environmental Protection Agency 2013). Up to 66% of air pollution-related deaths in India are the result of diesel vehicle exhaust emissions, which accounted for about 385,000 deaths in 2015, as noted by the *Economic Times* (2019). Therefore, it is vital to reduce this huge source of air pollution, and moving away from using individual private cars toward using shared vehicles could assist with fixing these issues. In this specific circumstance, carpooling has been a widely acknowledged idea to execute intelligent transport systems (ITS) in smart cities and to diminish the gas emanations, fuel use, and traffic management. Carpooling is one of the measures that citizens can embrace, and that organizations can implement for sustainable development.

BlaBlaCar, a company that dominates the carpooling market of many European countries, conducted research on a projected estimate of the reduction in carbon dioxide emissions by vehicles in the year 2023 (BlaBlaCar 2020). This is depicted in Figure 9.2. As demonstrated in the graph, BlaBlaLines is projected to save 1.88 million tons of carbon.

UberPool and OlaShare, two of the most popular carpooling services in India, have been successful in saving around 3.5 million liters of fuel by cutting short 74 million kilometers of travel, thereby reducing 8 million kilograms of carbon emissions (Jain 2017). Their services are available in most metropolitan cities, like Kolkata, Mumbai, New Delhi, and Chennai.

A study in the United States demonstrated that 1% of private vehicles joining carpooling can save three billion liters of gas and 10% can save 28 billion liters in one year (Kwon and Varaiya 2008). As expressed by the State of European Car-Sharing report, carpools show 15–25% fewer $CO_2$ discharges in comparison to private vehicles (MOMO 2010). A study by Seyedabrishami et al. (2012) states that

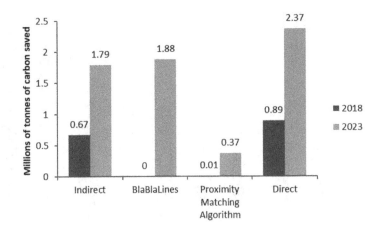

**FIGURE 9.2**   Projected $CO_2$ savings (BlaBlaCar 2020).

approximately 240 million liters of fuel can be saved every year if 30% of passengers started using carpooling in Tehran, Iran. The Environmental Assessment Report expresses that vehicle sharing can possibly diminish emissions by 40–50% (MOSES 2005).

## 9.3   GROWTH OF CARPOOLING OVER THE YEARS

Research has been conducted all over the world to realize and reap the economic, environmental, and social benefits of carpooling. Many countries have adopted ride-sharing techniques promoted by their governments or different private service providers. Carpooling or ride sharing started flourishing in the early 2010s, and by 2017, carpooling had over 10 million users worldwide. According to research conducted by Frost and Sullivan, the number of carpooling uses will cross 36 million by 2025, keeping up the yearly development pace of 16.4% (Frost and Sullivan 2016; Waszkowski 2020). It is estimated that the total number of carpool users across the world will increase from 11 million in 2017 to around 36 million by 2025, as shown in Figure 9.3 (Waszkowski 2020).

The ridesharing market surpassed USD 2.5 billion around 2019 and is ready to develop at a CAGR of above 24% between 2020 and 2026. Globally, governments are executing severe emission guidelines, advocating for the use of carpooling arrangements, and encouraging the market's development. The expanding traffic in urban regions combined with swamped public vehicle services is boosting the vehicle sharing business sector revenue. Workers that develop a reliance on carpooling administrations can be at least somewhat credited with the public transportation sector's rising security concerns (Wadhwani 2020). Figure 9.4 shows the projected increase in demand of carpooling services in the business and private sectors. A uniform growth is noticed in both these sectors.

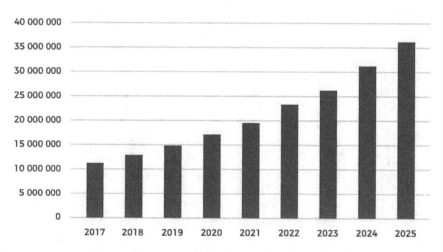

**FIGURE 9.3**   Projected number of users from 2017 to 2025 (Waszkowski 2020).

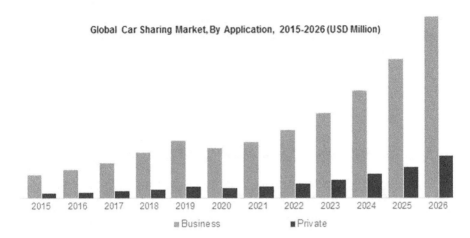

**FIGURE 9.4**  Increasing demand for car-sharing services in businesses (Wadhwani 2020).

The carpooling market is expected to develop further by providing additional services in currently serviced cities and expanding to new markets across the globe. By 2028, the carpooling market is expected to further increase through additional choices of vehicles, such as vans, electric cars, and luxury vehicles, and improved searching techniques for vehicle sharing systems (Gaille 2018).

At present, the primary rideshare mobility markets are dominated by Europe and the United States. By 2021, analysts speculate that Asia Pacific countries will gradually surpass the European and US domination in this sector, as shown in Figure 9.5. Figure 9.6 shows the share of market capture of various carpooling service providers in one such expanding Asian country, India.

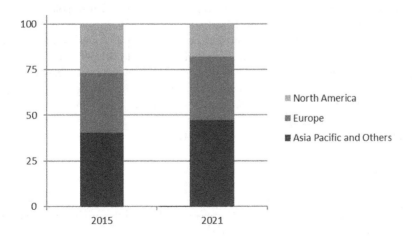

**FIGURE 9.5**  Breakdown of the carpooling market by global regions (Wagner 2017).

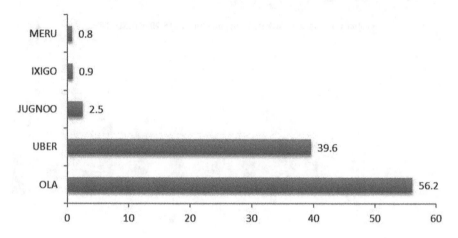

**FIGURE 9.6**  Different carpooling services in India and their market uses by percentage (Keelery 2018).

## 9.4  APPLICATION OF CARPOOLING IN SUSTAINABLE CITY DEVELOPMENT

Various governments, agencies, and organizations have launched many technology-aided initiatives to ensure the successful implementation of carpooling services to achieve sustainable development in the major cities. Several popular initiatives are explored in the following sections.

### 9.4.1  GROWSMARTER

To showcase 12 smart city solutions, GrowSmarter aims to develop city-level smart arrangements by implementing green and sustainable solutions in Cologne (Germany), Stockholm (Sweden), and Barcelona (Spain). These solutions range between cutting edge technology based on data and innovation, well-associated urban mobility using carpooling, and legitimately incorporating sustainable power sources into the city's supply. This initiative plans to lower pollution effects by: (1) decreasing the requirement for non-renewable energy, (2) decreasing ozone-depleting particle emissions by encouraging the use of renewable energy sources, and (3) diminishing vehicle discharges (GrowSmarter n.d.). GrowSmarter aims to use car sharing to lower the joint capital and operational costs of private vehicles, providing cost-effective economic growth.

### 9.4.2  CIVITAS

City Vitality Sustainability Initiative works toward a sustainable transportation framework. This initiative focuses on measures that favor the productive use of sustainable mobility systems, mainly carpooling. Presently, more than 200 cities across

Europe are members of this ridesharing network, including Debrecen (Hungary), Krakow (Poland), Rome (Italy), Perugia (Italy), Toulouse (France), and Stuttgart (Germany). From 2005 to 2007, around 1,600,000 km was saved in total, with a reduction of 340,000 kg of carbon dioxide emissions in these European cities (Parezanović et al. 2019).

### 9.4.3 GreenCar

GreenCar is an Indian non-profit initiative that works toward advancing carpooling, which will lead to a decreased use of fuel, result in lower pollutant emissions and congestion on road networks, increase green solutions, and promote sustainable development in urban areas (GreenCar n.d.). GreenCar decreased traffic and environmental pollution on the busy streets of India, a highly populated nation. Its vision focuses on limiting carbon dioxide emissions to cleanse the atmosphere. It is India's largest car-sharing non-government organization.

### 9.4.4 Green America

Green America is an initiative for a socially just and environmentally manageable society. The initiative focuses on three sectors for positive change by demanding social equity and ecological well-being: (1) renewable and pollution-free sources of energy, (2) sustainable food and agri-business, and (3) investing to facilitate economic development. Green America's "Carpool for the Climate and Community" promotes various ridesharing service providers to ensure sustainable development of green cities (Green America n.d.).

## 9.5 POPULAR CARPOOLING SERVICE PROVIDERS ACROSS THE WORLD

A detailed comparison has been performed for different carpooling organizations' various SWOT (strength, weakness, opportunities, and threats) analyses. In this section, some of the world's popular carpooling services are mentioned, along with their SWOT analysis.

### 9.5.1 Uber

Uber, also known as Uber Technologies, Inc., is a popular business venture that provides transportation, food conveyance, carpooling, and vehicles-for-hire services. It was initially based in the United States and now has services in over 900 urban cities worldwide. uberPOOL is their easy-to-use and reasonably priced carpooling offer that simultaneously matches users with riders who originate from a similar zone and heading to the same general area (Uber n.d.).

### 9.5.2 Ola Cabs

Ola Cabs is an India-based mobility service that provides carpooling services (OlaShare), along with vehicles for hire (Ola n.d.). Ola was initially developed by ANI Technologies Pvt. Ltd., Bangalore, India, and now its services are available in various cities in India.

### 9.5.3 Car2go

Car2go was initially a car rental company based in Germany, but is now owned by Daimler AG and provides carpooling services in many American and European cities (Car2go n.d.).

### 9.5.4 Zipcar

Zipcar is an American carpooling organization and an auxiliary of Avis Budget Group. It gives vehicle reservations to customers and bills hourly or daily. Members may need to pay a monthly/yearly enrollment and vehicle usage costs (Zipcar n.d.).

### 9.5.5 MSU Carpooling

With over 50,000 students, Michigan State University (MSU), USA, is one of the largest universities in the country. MSU launched various flagship projects to restrict emissions of ozone-depleting substances. This incorporated endeavor gives carpooling services to employees and students to constrain scope 3 emissions discharged by individuals driving to work using private cars (Kaplowitz and Slabosky 2018).

### 9.5.6 Stars

Shared mobility opporTunities And challenges foR European citieS (STARS) was launched in 2017 to propagate carpooling in European cities. Due to various available technological advancements, carpooling is easily implemented in Europe (STARS-H2020 n.d.).

### 9.5.7 Didi Chuxing

Didi Chuxing Technology Co., a Chinese company in Beijing, provides application-based transportation administrations, like taxi and private vehicle hiring, carpooling, and bike sharing services (Didi Chuxing Technology Co., n.d.).

### 9.5.8 Meru Cabs

Meru Cabs is a ridesharing company based in India. It provides app-based transportation services, like city rides, carpooling, airport transfers, and outstation travel, to individuals and corporate buildings in 24 cities across India (Meru Cabs n.d.).

Table 9.1, summarizes the SWOT for these above-mentioned carpool service providers.

**TABLE 9.1**
**Strengths, Weaknesses, Opportunities, and Threats (SWOT) for Carpool Service Providers**

| Carpool Service Provider | Strengths | Weaknesses | Opportunities | Threats |
|---|---|---|---|---|
| Uber (UberPool) | Brilliantly connects the target market with available transportation Vast coverage in over 100 countries Better reputation and brand awareness: provide competitive merits First-mover advantage | Operation protection Customers demonstrate low levels of loyalty Switching costs for customers are low; customers likely to prefer price advantage | Capturing the rural market Providing innovative schemes and incentives to capture new customers Introduction of modern technologies and amenities | Competitors in different regions. Legal problems |
| OLA (OlaShare) | First taxi aggregator service provider in India Top class administration Ranks first in India, after acquisition of taxi for sure Major research and development and online application Rapid awareness through media advertising Huge client base Venture capital has made OLA monetarily solid | Drivers' misconduct affects the brand image Running cost as well as infrastructural investments has been enormous even though there is a steady demand for the service. | Market is gigantic, thus the potential is high Advancing web infiltration along with smartphone clients Increasing disposable income Gradual move of customers toward comfort makes promising market Acquisition of smaller brands in the same field | Rising rivalry Presence of numerous firms in the same field Absence of clear government guidelines Absence of guidelines Lower customer loyalty |

*(Continued)*

**TABLE 9.1 (*Continued*)**

**Strengths, Weaknesses, Opportunities, and Threats (SWOT) for Carpool Service Providers**

| Carpool Service Provider | Strengths | Weaknesses | Opportunities | Threats |
|---|---|---|---|---|
| Car2go | Early-mover advantage; largest client base in the region<br>Greater availability of cars<br>Ownership of shielded activity rights from the government<br>Premium vehicles for quality clients<br>Use of Daimler's capital market and brand notoriety | Suitable for brief trips<br>Price rise during rush hours<br>Enrollment fees during registration<br>Constraints on service due to inadequate vehicle density<br>Marketing shortcomings due to varying client segment | Collaboration with public establishments like airport, train, and ferry services<br>Improved estimating models and adaption to clients' needs<br>Ingenious approaches: electric mobility and autonomous driving<br>Growth through customer consciousness and awareness | Early-mover preferences nullified by minimum client transfer costs<br>Opening of the market to external service providers<br>Constraint of city administration on expansion |
| Zipcar | Expanded its membership<br>Retained existing members<br>Cordial relationship with clients<br>High growth in exclusive class<br>User-friendly system | High investment in new sector entry<br>Thin margins, high expenses | High fuel costs will lead to greater vehicle sharing<br>Scope of increasing customer base | Competitors<br>Escalation in diesel cost will reduce margin |
| MSU carpooling | Promotes greener and cleaner campus with less vehicles<br>Large parking lot availability<br>Shared responsibilities<br>Lowers car maintenance costs | Driver and vehicle dependent<br>May have to share rides with unfamiliar people<br>Limitation of mobile connectivity for accessing facilities | Promotes socializing<br>Innovative clean campus technology<br>Solution to parking woes<br>Lowering of pollution. | Non-adherence to new system<br>Spread of contagious diseases<br>Hostility among co-travelers<br>May not find enough passengers |

**TABLE 9.1 (*Continued*)**
**Strengths, Weaknesses, Opportunities, and Threats (SWOT) for Carpool Service Providers**

| Carpool Service Provider | Strengths | Weaknesses | Opportunities | Threats |
|---|---|---|---|---|
| STARS | Oversees and secures the supply of vehicles being used. Fit to supply services to heavy density localities. Suited for unidirectional and long duration travel Market visibility | Needs to relocate vehicles Extra car overhead to adapt to pcak loads Terminals decrease the comfort of coverage Stations need extra running costs | Cooperation with government authorities for better service as part of ITS Spread in other cities Stations may be set up based on demand | Ride-hailing Public transport, car rentals may lower business revenue |
| Didi Chuxing | Robust brand visibility High-budget research and development $50billion, Valuation: backing from BAT, Apple Scale advantage. Market share 80% | Progressive withdrawal of driver incentives after reaching critical mass Dispatch algorithm based on the nearest taxi for faster response Drivers reject short travels | Electric cars, self-driving cars Worldwide spread | New competitors Market saturation Less loyalty in price sensitive market Ethical and social issues like women's safety |
| Meru cabs carpool | One of India's biggest radio taxi service providers Solid brand image Single-touch booking facility Fleet strength 10,000 radio cabs Tracking of car and user location through GPS Round the clock client care facility Links with banks for making payments | Financial problems Age of fleet Negative reviews on online media Complaints and strikes by drivers Penetration is poor across Indian cities | Additional car rental and ride-sharing services Expansion in unexplored regions Safety and security of clients | Rivalry from numerous competitors Driver union activities Unexpected government rules |

## 9.6  CONCLUSION

Carpooling has proved to be an effective solution for reducing air and noise pollution, traffic congestion on road networks, and expenditure associated with individuals using private vehicles. It is a sustainable approach toward traveling because sharing trips with different individuals instead of using private vehicles reduces carbon emissions and results in a huge decrease in fuel consumption. Various private and public organizations have started implementing carpooling as a successful solution to transportation problems, as it is environment friendly and promotes the international goal of a greener and cleaner earth. Carpooling, or ridesharing, also lowers the need for parking spaces in the urban cities and acts as a source of income for part- and full-time drivers, generating new jobs. Another major contribution is the financial savings to users, as the cost of traveling is shared among many people, reducing the cost of living. Along with financial and environmental benefits, ridesharing promotes social communication and helps in cooperative growth among the users.

In spite of all the above-mentioned benefits, there are a few threats that affect the carpooling market. First, there is a threat to the safety of travelers in some cities because they need to share rides with strangers. Second, maintaining social distancing is not possible to a large extent with carpooling, as seen during the COVID-19 pandemic. The concerns related to infectious disease can be minimized by the service providers if they practice proper sanitization techniques.

To ensure the most effective implementation of carpooling, there needs to be additional research conducted on a global scale. Various algorithms are being designed to help drivers and service providers make decisions based on scientific grounds. Recent advancements in the field of ICT aid in further optimizing the carpooling sector by using navigation tools and GPS tracking systems. By properly implementing carpooling and working toward removing the factors that are threats to this sector, major sustainable development solutions can be achieved through the implementation of ridesharing frameworks across the globe.

## REFERENCES

BlaBlaCar. 2020. "STUDY- Carpooling Saves More Than 1.6 Million Tonnes of $CO_2$ a Year, Whilst Doubling the Number of People Traveling. *BlaBlaCar.* https://blog.blablacar. com/newsroom/news-list/zeroemptyseats.

Cambridge Dictionary. n.d. "Carpooling." Accessed 2020. https://dictionary.cambridge.org/ dictionary/english/carpooling

Car2go. n.d. "Car Sharing in the U.S.A." Accessed 2020. https://www.car2go.com/US/en/?.

Collins Dictionary. n.d. "Definition of Car Pool." Accessed 2020. https://www.collinsdiction- ary.com/dictionary/english/car-pool.

Didi Chuxing. n.d. "DiDi: Official Website." Accessed 2020. https://www.didiglobal.com/.

*Economic Times.* 2019. "Majority of Air Pollution Deaths in India Linked to Diesel Vehicle Emissions: Study." *India Times.* https://economictimes.indiatimes.com/news/ politics-and-nation/majority-of-air-pollution-deaths-in-india-linked-to-diesel-vehicle- emissionsstudy/articleshow/ 68184315.cms.

Frost and Sullivan. 2016. "Future of Carsharing Market to 2025." Accessed 2020. https:// store.frost.com/future-of-carsharing-market-to-2025.html.

Gaille, B. 2018. "20 Car Sharing Industry Statistics, Trends & Analysis." *Brandongaille.com.* https://brandongaille.com/20-car-sharing-industry-statistics-trends-analysis.

Government of Canada. n.d. "Environment and Climate Change Canada." Accessed 2020. https://www.canada.ca/en/environment-climate-change.html.

Green America. n.d. "Carpool for the Climate and Community." Accessed 2020. https://www.greenamerica.org/green-living/carpool-climate-and-community.

GreenCar. n.d. "Carpooling, Cabpooling Terms & Conditions — GreenCar" Accessed 2020. https://www.greencar.ngo/terms.

GrowSmarter. n.d. "GrowSmarter: Smartcities Information System." Accessed 2020. https://smartcities-infosystem.eu/sites-projects/projects/growsmarter.

Jain, A. 2017. "Tech-Powered Carpooling Can Contribute to a Sustainable Mobility Future." *Shared Mobility.* https://www.sharedmobility.news/tech-powered-carpooling-can-contribute-sustainable-mobility-future/.

Kaplowitz, S., and A. Slabosky. 2018. "Trying to Increase Carpooling at a Major U.S. University: A Survey and an Intervention." *Sustainability: The Journal of Record* 11(2). https://doi.org/10.1089/sus.2017.0020.

Keelery, S. 2018. "Infographic: Ola at Full Throttle in India's Ride-Hailing Market." *Statista.* https://www.statista.com/chart/14650/ride-hailing-market-share-india/.

Kwon, J., and P. Varaiya. 2008. "Effectiveness of California's High Occupancy Vehicle (HOV) System." *Transportation Research Part C Emerging Technologies* 16(1): 98–115. doi: 10.1016/j.trc.2007.06.008.

Manzini, R., and A. Pareschi. 2012. "A Decision-Support System for the Carpooling Problem." *Journal of Transportation Technologies* 2(2). doi: 10.4236/jtts.2012.22011.

Meru Cabs. n.d. "Meru." Accessed 2020. https://www.meru.in/.

MOMO. 2010. "The State of European Car-Sharing." Final Report D 2.4 Work Package 2 Bundesverband CarSharing e. V. June 2010. http://www.car-sharing.info/images/stories/pdf_dateien/wp2_report__englisch_final_2.pdf

MOSES. 2005. "Mobility Services for Urban Sustainability." EUROPA: Moses Deliverable 6.2.

Mulders, C. 2012. *"Carpooling,* a Vehicle Routing Approach." Master's Thesis, Université catholique de Louvain.

Ola. n.d. "Ola Cabs." Accessed 2020. https://www.olacabs.com/.

Parezanović, T., M. Petrović, N. Bojković, and Prof. S. Pejčić Tarle. 2019. "Carpooling as a Measure for Achieving Sustainable Urban Mobility: European Good Practice Examples." XXIII International Scientific-Technical Conference "trans & MOTAUTO '15."

Rijavec, R., N. Dadashzadeh, M. Žura, and R. Marsetič. 2020. "Park and Pool Lots' Impact on Promoting Shared Mobility and Carpooling on Highways: The Case of Slovenia." *Sustainability* 12(8): 3188. doi: 10.3390/su12083188.

Seyedabrishami, S., A. Mamdoohi, A. Barzegar, and S. Hasanpour. 2012. "Impact of Carpooling on Fuel Saving in Urban Transportation: Case Study of Tehran." *Procedia — Social and Behavioral Sciences* 54: 323–331. doi: 10.1016/j.sbspro.2012.09.751.

STARS-H2020. 2020. "Shared Mobility Opportunities and Challenges for European Cities." Accessed 2020. http://stars-h2020.eu/.

Uber. n.d. "Het-Platform Verkennen." Accessed 2020. https://www.uber.com.

United States Environmental Protection Agency. 2013. "US Transportation Sector Greenhouse Gas Emissions: 1990–2011." *United States Environmental Protection Agency, Office of Transportation and Air Quality*, EPA-420-F-13-033a.

Wadhwani, P. 2020. "Car Sharing Market Size by Model (P2P, Station-Based, Free-Floating), By Business Model (Round Trip, One Way), By Application (Business, Private), Industry Analysis Report, Regional Outlook, Application Potential, Price Trend, Competitive Market Share & Forecast, 2020–2026." *Global Market Insights, Inc.* https://www.gminsights.com/industry-analysis/carsharing-market.

Wagner, I. 2017. "Worldwide — Car Sharing Market by Region 2021." *Statista*. https://www. statista.com/statistics/777157/global-car-sharing-market-breakdown-by-region/.

Waszkowski, E. 2020. "Car Sharing and Transportation Trends (Updated)." *Future Mind*. https://www.futuremind.com/blog/car-sharing-and-transportation-trends.

ZipCar. n.d., "Car Sharing: An Alternative to Car Rentals," Accessed 2020. https://www. zipcar.com/

# 10 Learning the Scandinavian Approach to Green Infrastructure, Energy, Circular Economy

*Peyush Karel and Xavier Savarimuthu, SJ*

## CONTENTS

## 10.1 INTRODUCTION

Sustainability has been defined by the United Nations Brundtland Commission as "meeting the needs of the present without compromising the ability of future generations to meet their own needs."

The Scandinavian countries of Sweden, Norway, Denmark, Iceland, and Finland contribute a small percentage to the world gross domestic product (GDP) but have a lower ecological footprint compared to most of the nations. They have portrayed unprecedented progress in the transition toward a greener society and as seen in Table 10.1, they have been placed among the top 20 green countries in terms of Environmental Performance Index (EPI) scores consistently (2020 results, Environmental Performance Index).

**TABLE 10.1**

**2020 Environmental Performance Index Rankings and Scores of Nordic Countries**

| Country | EPI Ranking | Environmental Performance Index | Environmental Health | Ecosystem Vitality |
|---------|-------------|--------------------------------|---------------------|-------------------|
| Denmark | 01 | 82.50 | 91.70 | 76.40 |
| Finland | 07 | 78.90 | 99.30 | 65.30 |
| Sweden | 08 | 78.70 | 98.40 | 65.60 |
| Norway | 09 | 77.70 | 98.50 | 63.80 |
| Iceland | 17 | 72.30 | 98.10 | 55.00 |

*Source:* Data from the Environmental Performance Index, Results.

Sweden became the first country to establish an Environment Protection Agency back in 1967 and was instrumental in the creation of the United Nations Environmental Programme (UNEP). The Nordic countries have various agreements and frequent reporting mechanisms in place to reach their sustainability targets. Generation 2030 was adopted to achieve the UN's Agenda 2030 (Nordic Council of Ministers 2017). Also, the Committee for a Sustainable Nordic Region was established to cater to issues relating to environmental degradation.

The Scandinavian nations have been influential in the formulation of the Paris climate agreement and have ranked among the world's top innovators in Global CleanTech Innovation Index (World Intellectual Property Organization [WIPO] 2019). While there are no all-encompassing blueprints for environmental sustainability, a conscious, indigenous, and innovative adaptation of Nordic measures in infrastructure, energy, and waste management can go a long way toward environmental protection and restoration.

## 10.2 INFRASTRUCTURE

The multitude of construction types has contributed to unprecedented degradation and harm to the environment with their use of environmentally degrading materials, inputs, and unsustainable operating systems. Buildings and constructions around the world account for one-third of global energy consumption and contribute to 40% of the world's total $CO_2$ emissions (International Energy Agency 2020). Utilizing about 40% of the world's raw materials, buildings contribute to environmental degradation through associated activities like raw material extraction, production, and transportation. Demolition further deteriorates the environment by producing waste which gets dumped in landfills or incinerators (down-cycling of waste results in high loss of material value) polluting both air and soil. The construction sector plays a key role in the creation of sustainable economies. A transition toward a circular economy is not restricted to plastics or circular electrical and electronic equipment (CEEE) but extends to include refurbishing and reuse of construction materials, energy, and waste generated through demolition as well.

In this regard, the Environment and Economics Group (MEG) under the Nordic Council of Ministers (NCM) has identified various policy instruments including building information modeling (BIM) system that promote a transition toward the concept of a circular economy and achieve resource efficiency in this sector and have encouraged a lot of private players to shift their focus onto this subject (Anna Kortesoja et al. 2018).

### 10.2.1 RAINWATER HARVESTING: THE SOUL OF NØRREBRO, DENMARK

This is a project created for Nørrebro, the City of Copenhagen as the client (Hans Tavsens Park and Korsgade), and aims at establishing new ecosystems to reduce flooding and creating a microcosm reflective of diverse cultures to address the diverse needs and avail opportunities for urban living in the arena. The project (Designer: SLA) recognizes water from heavy cloudbursts or rainfall as a quintessential resource and captures water by creating a synergy between the three complementary cycles, namely the hydro-geological cycle, the biological cycle, and the social cycle.

As seen in Figure 10.1, the project, post-implementation, aims to transform the park into a multitude of rainwater catchment basins, purify the resource through biotopes (city park greenery) and facilitate the flow of collected rainwater to the Copenhagen lakes (One planet 2017). Tanks would be placed to collect water from rooftops all over the place such that residents, schools, and municipalities can use the

**FIGURE 10.1** Hans Tavsens Park. (Rendering credit: Beauty and the Bit.)

collected water for various purposes. The Hans Tavsens Park would be fitted with a retention volume of 18,000 m³ to maximize retention and usage.

Having won the prestigious Nordic Built Cities Challenge (Nordic Innovation 2016), the project also aims to involve schools (The Blågård School, Nørrebro Park School, HTØ and HTV, Et Frie Gymnasium) by providing children with project-related knowledge such that they can improvise and maintain the project and its vision in the near future.

## 10.2.2 Timber-Based Element Systems: PuuMera, Vantaa

Timber based elemental systems (TES) are viable and environment-friendly substitutes to BEC (prefabricated concrete elements) which lead to high energy consumption in operations and low rates of material reusability post demolition. A mega construction project called "PuuMera" (Builder: Rakennusliike Reponen) in Vantaa, Finland uses laminated timber structures along with recycled glass and untreated wood which can reduce the energy consumption levels of a building similar to those of a passive house (Passive House Institute (PHI): a passive house is a building standard that is truly energy-efficient, comfortable, and affordable).

With a surface area of 10,120 square meters (m²), the U-value (thermal transmittance) of the roof 0.08 W/m²K (Watts-per-Metre-Square-Kelvin) and of the walls 0.12 W/m²K, the building generates electricity through solar panels fitted on the rooftop and is a passive building in terms of energy consumption as defined by the VTT Technical Research Centre of Finland. As seen in Figure 10.2, the external

FIGURE 10.2 PuuMera building in Vantaa. (Photo credit: VVR.)

walls of the building are constructed with laminated timber while both the beams and the partitions endorse cross-frames (One planet 2017). PuuMera demonstrates that construction using backlit strategies can take place in a cost-efficient manner toward the creation of an energy-efficient society.

### 10.2.3 CLIMATE-SMART TECHNOLOGY: GREENHOUSE, AUGUSTENBORG

The introduction of innovative climate-smart concepts for buildings reduces the overall energy consumption levels of the buildings and makes them energy efficient through a shift to renewable energy sources. Sweden has taken responsibility for the development of Augustenborg on sustainable lines and an example backing the change is a sustainable housing project called "Greenhouse Augustenborg" in Malmö, Sweden (Builder: NCC Sweden).

The building consists of a total of 56 apartments and allows residents to undertake urban gardening in spaces provided by the semi-open balconies. Figure 10.3 shows that the building is fitted with a greenhouse on the rooftop, and allows the residents to cultivate plants on a sharing basis. The reading panes mounted on walls of each apartment allow residents to take control by providing accurate data about their energy consumption, including data on water and green energy (produced by the solar cells) consumption along with information on bike rentals to promote green transportation. The panes also educate the residents on cultivation, benefits of organic food, and conserving energy, thereby, creating environmental consciousness among its residents using smart technology (Alexander Sandstedt 2016).

**FIGURE 10.3** Greenhouse in Augustenborg built by NCC. (Photo credit: Klas Andersson/ NCC.)

As per NCC Sweden, the project has been environmentally certified as "Environmental Building Gold" — the highest mark for homes with low energy needs and a healthy indoor environment.

## 10.3   ENERGY

With expanding energy inclusivity, the primary energy consumption of the world recorded the fastest growth of about 2.9% (since 2010) in the year 2018. Also, the carbon emissions recorded an increase of 2.0% (0.6 gigatons), the highest in the last seven years (BP plc 2019).

Even with increased commitments to a transition toward green energy, and renewable energy registering a growth of 14.5%, almost 45% of the increase in energy consumption is attributed to an increase in natural gas demand (International Energy Agency 2019). This has raised concerns for economies globally and has called for large scale unitary activism toward the development of clean energy sources. Innovating the concepts of district heating, heat pumps, and fuel cell technology and harnessing the potential of human energy or body heat, the Nordic countries have developed projects that have become learning tools for the global economy in the development of a carbon-neutral energy system.

The Nordic nations have registered a high share in renewable energy consumption. Sweden, Iceland, and Norway consumed 65.8%, 93.1%, and 106.4% renewable energy as a share of gross electricity consumption in 2015, respectively (Eurostat 2017). Sweden's share in renewable energy consumption as a percentage to its gross final energy consumption stood at 54.5% in 2017, the highest among EU nations (Eurostat 2019), followed by Finland (41%).

### 10.3.1   DISTRICT HEATING AND CHP

Owing to the harsh climate conditions, more than half of the EU's total energy is consumed by the heating and cooling sector. The fossil fuels act as primary energy providers for heating and cooling applications while renewable energy sources contribute a minuscule percentage. Denmark, however, accounted for over 42% of district heat generated using renewable heat in 2014 (Martin Christoph Soini et al. 2017).

District heating is an innovative system for disseminating heat to a locality through underground insulation pipes produced by a heat cogeneration plant located in a centralized area. Figure 10.4 shows Oersted gas-fired power station supplying heat to Copenhagen. Geothermal heating, solar heating, heat pumps, biomass heating, and heat waste from nuclear power generation are some of the methods employed by the centralized boiler to heat and store water.

Combined heat and power (CHP) systems generate electricity by utilizing a fuel source and are often employed at the district heating facilities. The electricity generated is sold to the power grids and heat waste is transferred to heating plants for circulation in the underground insulation pipes that cater to the heating needs of the locality. The heat waste generated by the households is transferred back to the CHP systems leading to new electricity generation which improves overall energy efficiency by 40–45%, making the entire process sustainable (Jenni Patronen et al. 2017).

**FIGURE 10.4**    Oersted gas-fired power station supplying heat to Copenhagen.

Industrial waste heat from various manufacturing plants or data centers can act as a primary fuel source for district heating and CHP systems. Low-grade industrial heat is discharged into the environment after poor industrial processing and utilization of such waste heat in district heating can lead to high energy recovery rates and further the process of green energy production.

## 10.3.2   BODY HEAT

The heat released from individuals in the conduct of various activities is known as body heat. This heat gets released in the atmosphere in the form of thermal energy. Governments have tried harnessing this heat but have had little success. For instance, the Mall of America in Minnesota, USA, utilizes the heat released from the shoppers to keep the mall warm in winter.

Sweden has found success in harnessing the application potential of body heat by capturing and transferring the heat to other buildings. One fine example of Swedish innovation is the model developed by Jernhusen, a Stockholm real estate company that transfers heat captured from commuters in Stockholm's Central Station to the office building of Kungbrohuset located 100 yards away (Kaushik Patowary 2019).

The company has installed heat exchangers in Central Station's ventilation system which captures the excess body heat released from the commuters as well as from the kitchens of the restaurants located in the station and uses it to heat the water in the underground tanks. The warm water is then channeled to the "Kungbrohuset" building through a system of underground pipes becoming a part of the overall heating system. This has resulted in reductions of about 25% in energy costs for the building (Tara Kelly 2010).

The novel approach to energy utilization is, however, primarily dependent on certain factors that dictate the overall efficiency and effectiveness of the approach;

they include the overall energy consumed by the pumping mechanism, regular high-density of commuters to generate enough energy, and shorter distance between the energy source and the transferee building. The mechanism is being adopted by governments and innovators across the globe and can lead to a switch to unconventional greener and newer energy options in the years to come.

### 10.3.3 HYDROGEN FUEL CELLS

As per the corporation Hydrogenics, "A fuel cell is a device that converts chemical potential energy (energy stored in molecular bonds) into electrical energy. A PEM (Proton Exchange Membrane) cell uses hydrogen gas ($H_2$) and oxygen gas ($O_2$) as fuel. The products of the reaction in the cell are water, electricity, and heat" (Hydrogenics 2020).

Hydrogen fuel cells are a breakthrough in power innovation and are high energy-efficient (energy content: 55%) replacements to fossil fuels in power production and consumption. Unlike batteries, they need not be recharged but they need a continuous supply of fuel and oxygen, each of which is found in abundance. Fuel cells are scalable (can be stacked) and operate silently, owing to no moving elements, and have huge application potential in areas like transport, data centers, and power backup systems. Along with the European Union's (EU) Fuel Cells and Hydrogen Joint Undertaking (FCH JU), the Nordic countries have been recognized as forerunners in the development of this technology. The EU's Hydrogen Mobility Europe (H2ME) initiative and the Second Joint Initiative for Hydrogen Vehicles across Europe (JIVE 2) project are working to deploy hydrogen refueling stations and hydrogen-powered buses within the Nordic countries of Sweden, Norway, Iceland, and Denmark.

In 2013, Copenhagen became the first city to inaugurate hydrogen fuel cell vehicles and since then, the country of Denmark has been taking active steps toward expansion of hydrogen-powered vehicle base with current projects (Copenhagen Capacity) expecting to deploy 200 hydrogen buses by 2020. Nel ASA, an organization in Norway, has helped in the development of hydrogen fuel cells and the construction of refueling stations across the Nordic region. Figure 10.5 shows a hydrogen fuel dispenser. The company suggests that it will expand the capacity of the plant in Notodden, Norway to 360 MW per year, the largest in the world (Jackson Carr 2019).

Sweden is transforming the hydrogen cell industry with FCHEA member myFC, the leading company in manufacturing micro fuel cells. In collaboration with Gränges AB, myFC is working on its product LAMINA REX which will enable onboard hydrogen generation and double the range of medium-sized electric vehicles (FuelCellsWorks 2019).

Hydrogen technology has found application in powering housing complexes too with Sweden building an energy-sufficient complex in its Vårgårda Municipality by Vårgårda Bostäder in collaboration with Nilsson Energy and Better Energy. The project is expected to rely only on hydrogen fuel cells and solar photovoltaics for energy generation and storage. The combination of solar photovoltaics caters the huge energy input requirements of hydrogen cells and minimizes energy waste. With an installed generation capacity of 659 kWp (kilowatt "peak"), the solar energy will power the 172 apartments in the complex (Better Energy 2019).

**FIGURE 10.5**   Hydrogen fuel dispenser.

Hydrogen power generation, however, is still in its nascent stages and with by-products as water and heat, they have major potential in helping governments achieve their zero-emission targets.

## 10.4   WASTE

The cities all around the world generated an aggregate 2.01 billion tons of solid waste in the year 2016, amounting to a footprint of 0.74 kilograms per person per day (World Bank 2019). The rate of waste generation, however, is increasing because of ever-increasing consumption and mass scale urbanization. 68% of the world population is expected to migrate to the urban areas, adding 2.5 billion people to the present urban population of the world (United Nations Department of Economic and Social Affairs 2018). The anticipated increase in waste generated throughout the world is about 70%, that is, to 3.40 billion tons by 2050. Also, the United Nations Food and Agriculture Organization (FAO) reports that one-third of all food is spoiled or squandered globally, that is, a loss of $940 billion (850 billion euros) when quantified in financial terms. It also generates 8% of the global greenhouse gas emissions (Food and Agriculture Organization).

While countries have started to adjust to the new scope of recycling activities, no contributions (solitary or unified) have been significant or at par with the crisis. Landfills account for about 59% of global waste (Derek Thompson 2012) and lead to soil and groundwater contamination by releasing hazardous toxins and greenhouse gases.

According to data published by Eurostat, countries like Denmark, Norway, and Iceland produce 777 kg, 754 kg, and 656 kg per person of municipal waste respectively, and rank in single digits in terms of waste generation (Marta Martinez and Cristina Matamoros 2018). However, the Scandinavian countries have set examples for governments all over the world with their efficient and effective waste disposal mechanisms and ambitious targets such that not only have they been able to manage waste produced in their regions but have also started importing waste from other countries.

The Nordic countries have been working together in a system through forums and councils, like "Nordic programme to reduce the environmental impact of plastic 2018," to promote economies based on longevity of plastic product life cycles, extended producer responsibilities, and designs for toxin-free products and recycling.

Shift in focus from waste electrical and electronic equipment (WEEE) to CEEE to design a circular system that enables reprocessing of materials, in multiple cycles, back to the same product has helped in inventory reduction and efficient reusability of industrial products.

## 10.4.1 Panta Recycling System

"Panta" refers to handing something over in return for money. It is a system of recycling in which individuals pay a refundable deposit for single-use drink containers. The deposits differ with the geography of states as well as with the reusability and recyclability of products. When these containers, including aluminum cans and PET bottles, are deposited in the reverse vending machines placed at various locations, such as shopping markets, the individuals get refunded for making that deposit. The machines generate a receipt that people encash at the till. The recycled bottles and cans get transported to recycling centers in the vicinity after being deposited into the panta machines.

The system is run by companies: Returpack Svenska AB (in Sweden) and Infinitum (in Norway). Sweden recycled 84.9% of its aluminum cans and PET bottles via the panta system in 2016, that is, 18,356 tons of aluminum and 22,361 tons of PET material in one year (TheLocalse 2018). Norway recycles about 97% of total plastic bottles with this panta system of recycling through a company called Infinitum (Matthew Taylor 2018).

## 10.4.2 Stationary Pneumatic Refuse Collection

Stationary pneumatic refuse collection (SPRC) refers to a system in which refuse is transported to long distances (underground) through pipes to a group station where it is compacted in closed containers. Figure 10.6 depicts the functioning of a SPRC. Air is employed to move the waste in a stationary pneumatic refuse assortment system. Within the pipe system, exhausters produce a negative pressure that successively generates the airflow. Once air enters the underground pipes at gas pressure, it entraps the solid waste and conveys it to the gathering station (Envac).

SPRC ends up in circumstantial and sleek handling of multiple types of waste. Every separate waste stream is backed by one refuse such that two to four separate waste streams can be handled using the same transport pipe network. Within the

**FIGURE 10.6**   Working of a SPRC. (Rendering credit: Envac Group.)

assortment station, every waste stream is directed toward a designated container. By grouping every waste kind singularly, the system ensures that waste and recyclables do not get mixed in the process. Primarily introduced in Sweden, the SPRC system has currently been adopted by multiple countries, each with their own modifications (IVL Swedish Environmental Research Institute).

### 10.4.3   PRODUCT DATABASE AND COMMON STANDARDS

In Sweden, a company named Standard Solutions Group AB provides SSG Product Database service which helps in the organization and structuring of spare parts produced and stocked by various industrial manufacturers and suppliers registered for the service. The SSG Product Database handles all part maintenance by providing a description to spare parts and matching them against existing database inventory. Unique SSG article numbers are assigned to new parts to enhance their identification and the entire inventory is visible and searchable online such that affiliated plants can view their stock and even borrow from other plants if the need be. The database offers unique transparency which makes it possible for plants to reduce capital blockage, streamline purchasing costs, increase plant availability, improve warehousing maintenance processes, enhance productivity, and cut administration.

Also, SSG draws uniform standards for industrial processes and systems that stress upon extended producer responsibilities and a lifecycle economy. Such uniformity enhances comparison between products and new manufacturers and suppliers are required to meet the standards to share a piece of the procurement process.

The aim of such standards is the recognition and satisfaction of needs and the creation of energy efficiencies by utilization of minimal resources possible in the process of production. The establishment of common standards within industries leads to increased operational reliability, personal safety, and thus increased sustainability. At this time, SSG Standard Solutions Group has drawn up more than 450 technical standards.

## 10.5 CONCLUSION

The metamorphosis toward sustainability, as depicted by the Nordic states, has proven to be highly significant in terms of contributing toward a realizable vision for the planet. The progress made by Scandinavian countries in transition toward green and sustainable construction and creation of passive houses, development of fuel cell technology and harnessing body heat, management of industrial waste and rewarding recycling, set examples for innovators and governments across the world toward reducing the environmental footprint and still catering to the increasing rates of urbanization. The green mechanisms innovated by these nations serve as models waiting to be emulated by other members of the globe.

The Nordic countries have certain urban plans in place that make sustainability a part of daily activities; a unique manner of living that has developed equally within the lifestyles of the Nordic population (Tine Sundtoft 2018). As seen in Table 10.2, the countries rank among the top 12 nations across the globe in the quality of living index 2020.

Influenced by increasing interest in Nordic urban planning, the region's three universities: Malmö University (Sweden), Arctic University (Norway), and Roskilde University (Denmark) have collaborated and designed an international master's program to provide lessons to the globe on the facts and plans of Nordic countries to promote sustainability and livability (Maddy Savage 2019).

Strategies for economic and ecological restoration have led to the advent of environmental, social, and governance (ESG) investing, also known as "responsible investing" with an increasing number of investors' interest in firms that have models in place to reduce their carbon footprints. Such investing has paved the way for various investment firms toward the adoption of stricter codes and standards for the choice of holdings and exclusion. Norway's sovereign wealth fund managed by NBIM, Norway's pension fund KLP, Finnish pension fund Varma, Sweden's AP and DNB Asset Management are some of the firms that have imposed absolute ceilings for volume and revenue limits on carbon emissions and income arising out of fossil fuel usage respectively (Al Jazeera 2020).

**TABLE 10.2**
**Ranks of Nordic Nations in Quality of Living Index 2020**

| Country | Rank | Quality of Life Index |
|---------|------|----------------------|
| Denmark | 01 | 192.67 |
| Finland | 03 | 190.22 |
| Iceland | 07 | 181.75 |
| Sweden | 11 | 175.95 |
| Norway | 12 | 175.19 |

*Source:* Data from Numbeo, Quality of Life Index by Country 2020 Mid-Year.

Such innovation toward building environmental conscience in all sectors governing global economies is key to sustainability.

The fact is, however, only a minuscule amount of impact has been generated in comparison to the rapid pace of environmental degradation, and there lies a huge potential for research and application to avoid the impending doom. The multiple aspects of sustainability desire initiatives requiring a wealth of cooperation between diverse international partners, including government agencies and private organizations, and generating consciousness and drive among those partners toward common action.

## REFERENCES

Al Jazeera. Reuters. 2020. Norway's sovereign fund takes significant steps against emissions. https://www.aljazeera.com/ajimpact/norway-sovereign-fund-takes-significant-steps-emissions-200525085653293.html.

Alexander Sandstedt. NCC. 2016. An ecological elevator in the eco-town of Augustenborg. https://www.ncc.se/vara-projekt/greenhouse-augustenborg/.

Anna Kortesoja, Marika Bröckl, Håkan Jonsson, Venla Kontiokari and Mikko Halonen. Nordic Council of Ministers. 2018. Nordic Best Practices: Relevant for UNEP 10YFP on sustainable buildings & construction and sustainable food systems. http://norden.diva-portal.org/smash/get/diva2:1177169/FULLTEXT01.pdf.

Better Energy. Off Grid Energy Independence. 2019. World's first energy self-sufficient housing complex. https://www.offgridenergyindependence.com/articles/16168/worlds-first-energy-self-sufficient-housing-complex.

BP plc. 2019. BP Statistical Review of World Energy 2019. Pages: 2–4. https://www.bp.com/content/dam/bp/business-sites/en/global/corporate/pdfs/energy-economics/statistical-review/bp-stats-review-2019-full-report.pdf.

Derek Thompson. The Atlantic. 2012. 2.6 Trillion pounds of garbage: where does the world's trash go? https://www.theatlantic.com/business/archive/2012/06/26-trillion-pounds-of-garbage-where-does-the-worlds-trash-go/258234/.

Envac. 2020. Envac in the city. https://www.envacgroup.com/waste-collection-reimagined/envac-in-the-city/.

Eurostat. 2017. Energy, transport and environment indicators. Page 72. Table 2.6.4. https://ec.europa.eu/eurostat/documents/3217494/8435375/KS-DK-17-001-EN-N.pdf/18d1ecfd-acd8-4390-ade6-e1f858d746da.

Eurostat. 2019. Energy, transport and environment statistics. Page 47. https://ec.europa.eu/eurostat/documents/3217494/10165279/KS-DK-19-001-EN-N.pdf/76651a29-b817-eed4-f9f2-92bf692e1ed9.

Food and Agriculture Organization (FAO) of the United Nations. 2015. Food wastage footprint & climate change. http://www.fao.org/3/a-bb144e.pdf.

FuelCellsWorks. 2019. myFC in joint development project for LAMINA REX with Gränges AB. https://fuelcellsworks.com/news/myfc-in-joint-development-project-for-lamina-rex-with-granges-ab/

Hydrogenics. 2020. Fuel Cells. https://www.hydrogenics.com/technology-resources/hydrogen-technology/fuel-cells/.

International Energy Agency. 2019. Global energy demand rose by 2.3% in 2018, its fastest pace in the last decade. https://www.iea.org/news/global-energy-demand-rose-by-23-in-2018-its-fastest-pace-in-the-last-decade.

International Energy Agency. 2020. Buildings: a source of enormous untapped efficiency potential. https://www.iea.org/topics/buildings.

IVL Swedish Environmental Research Institute. Smart City Sweden. 2020. Underground waste management. https://smartcitysweden.com/best-practice/8/underground-waste-management/.

Jackson Carr. FCHEA Intern. 2019. Scandinavia fuel cell industry developments. http://www.fchea.org/in-transition/2019/5/6/scandinavia-fuel-cell-industry-developments.

Jenni Patronen, Eeva Kaura and Cathrine Torvestad. Nordic Council of Ministers. 2017. Nordic heating and cooling: Nordic approach to EU's heating and cooling strategy. https://norden.diva-portal.org/smash/get/diva2:1098961/FULLTEXT01.pdf.

Kaushik Patowary. Amusing Planet. 2019. The building that steals your body heat. https://www.amusingplanet.com/2019/03/the-building-that-steals-your-body-heat.html.

Maddy Savage. Worklife. BBC. 2019. What the Nordic nations can teach us about liveable cities. https://www.bbc.com/worklife/article/20191112-what-the-nordic-nations-can-teach-us-about-liveable-cities.

Marta Rodriguez Martinez and Cristina Abellan-Matamoros. Euronews. 2018. Why do Scandinavians generate more waste than other Europeans? https://www.euronews.com/2018/01/30/why-do-scandinavians-generate-more-waste-than-other-europeans-.

Martin Christoph Soini, Meinrad Christophe Bürer, David Parra Mendoza, Martin Kumar Patel (UNIGE), Jasper Rigter and Deger Saygin (IRENA). International Renewable Energy Agency (IRENA). 2017. Renewable energy in district heating and cooling: a sector roadmap for remap. Page – 4. http://www.ourenergypolicy.org/wp-content/uploads/2017/03/IRENA_REmap_DHC_Report_2017.pdf.

Matthew Taylor. Guardian. 2018. Can Norway help us solve the plastic crisis, one bottle at a time? https://www.theguardian.com/environment/2018/jul/12/can-norway-help-us-solve-the-plastic-crisis-one-bottle-at-a-time.

Nordic Council of Ministers. 2017. Generation 2030. http://norden.diva-portal.org/smash/get/diva2:1153406/FULLTEXT01.pdf.

Nordic Innovation. 2016. Nordic Built Cities Challenge. https://www.nordicinnovation.org/programs/nordic-built-cities-challenge.

One planet. 2017. PuuMera – large-scale residential construction with wood. https://www.oneplanetnetwork.org/initiative/puumera-large-scale-residential-construction-wood.

One planet. 2017. The Soul of Nørrebro. https://www.oneplanetnetwork.org/initiative/soul-norrebro.

SSG Standard Solutions Group AB. 2020. Standardised structure and organisation. https://www.ssg.se/en/our-services/services/ssg-product-database/.

Tara Kelly. TIME. 2010. Body heat: Sweden's new green energy source. http://content.time.com/time/health/article/0,8599,1981919,00.html.

The Local se. 2018. That's pant! The story behind Sweden's bottle recycling scheme. https://www.thelocal.se/20180328/thats-pant-the-story-behind-swedens-bottle-recycling-system.

Tine Sundtoft. Nordic Council of Ministers. 2018. The Nordic countries in the green transition – more than just neighbours: Strategic recommendations for Nordic co-operation on the environment and climate in the run-up to 2030. https://norden.diva-portal.org/smash/get/diva2:1197618/FULLTEXT01.pdf.

United Nations Department of Economic and Social Affairs (UN DESA). 2018. 68% of the world population projected to live in urban areas by 2050, says UN. https://www.un.org/development/desa/en/news/population/2018-revision-of-world-urbanization-prospects.html.

World Bank. 2019. Solid Waste Management. https://www.worldbank.org/en/topic/urbandevelopment/brief/solid-waste-management.

World Intellectual Property Organization (WIPO). 2019. Global Innovation Index 2019. https://www.wipo.int/pressroom/en/articles/2019/article_0008.html.

# 11 Understanding the Economic and Insurance Industry Implications of the Australian Bushfires of 2019–2020

*Peyush Karel and Xavier Savarimuthu, SJ*

## CONTENTS

## 11.1 BACKGROUND AND INTRODUCTION

In this era of technological disruptions and scientific discoveries, a complex interplay of various macro and micro factors, often interrelated to one another in a multitude of ways, offers a more disciplined understanding and analysis of business and natural environments.

Bushfires refer to incidents of fire occurring on wild areas of land, typically bushes, scrubs, or forests. Such fires happen to take place on dry vegetation and spread rapidly over vast regions and thus, are difficult to control (National Geographic Kids 2020).

Australia is a part of a climate zone characterized by precipitation that is sufficient to allow vegetation to flourish over periods along with elongated periods of extremely dry and warm weather conditions (Munich RE 2020b). Under such conditions, the moisture content of plants frequently falls to dangerously low levels and turns them highly flammable. Figure 11.1 shows a fire danger ratings road sign of the Shire of Kondinin indicating catastrophic level of fire danger in the area.

Several elements like humidity, precipitation, temperature, and wind intensity affect bushfire seasons and their behavior (Understand Insurance 2020). For

**FIGURE 11.1**    Fire danger ratings road sign.

southern Australia, bushfires are a major threat in the dry summer months while in Northern Australia, the threat is in winter.

Australia has witnessed bushfires of unparalleled intensity over the past decades that have caused mass destruction. The first bushfire ever recorded in Australia was The Black Thursday in 1851, which burned over 5 million hectares of land in Victoria and claimed the lives of 12 people and thousands of cattle. The 2009 Black Saturday Bushfires led to a tragedy of an unrivalled scale. Asserted to be the deadliest bushfire to date, the 2009 Black Saturday Bushfires claimed 173 human lives, injured about 400 people, burned 4,00,000 hectares of land, and demolished more than 2,000 homes in over 78 communities nationwide (Stevens 2019). As estimated by Moody's Analytics, the 2009 bushfires were the costliest bushfires in Australia, with economic damages setting a record of about $4.4 billion and the insurance costs standing at $3 billion (Butler 2020).

The 2019 bushfires occurred on a scale unpredicted by any of the available models, and analysts claim that these bushfires set records, in terms of damages and costs, that far exceed those set by the 2009 bushfire crisis. The bushfire season began in early September 2019 and continued to blaze till April 2020. The fires covered a span of about 14 million acres in Southern Australia, with heavy impacts in Adelaide Hills, Kangaroo Island, and parts of the Australian Capital Territory. It also burned areas in Queensland and Tasmania. This bushfire season witnessed the death of over a billion animals, forcing many species toward extinction.

Early estimates suggest that the 2019–2020 bushfires have cost about $3.5 billion to the Australian economy, and Insurance Australia Group Limited (IAG) has estimated its insurance claims from natural disasters to be about $279 million for the first fiscal half of 2020. The following study offers a deeper analysis of the devastating crisis and its impacts.

## 11.2 CLIMATE CHANGE: A REASON?

Australia, one of the biggest carbon emitters per person, contributed to 1.4% of global $CO_2$ emissions from the domestic use of fossil fuels in 2017. Despite housing a small population of only 25 million, Australia's fossil fuel exports totaled its global carbon footprint to about 5%, making the country the world's fifth biggest emitter of carbon (Spence 2020).

Already host to ten of the biggest coal mines worldwide, Australia is considering operating one in Adani in Queensland, which would double the country's coal-based carbon emissions.

The National Disaster Risk Framework, along with 23 former fire chiefs and few emergency leaders, warned the authorities of impending threats of natural hazards occurring at unimagined scales due to climate change, and an urgent need for resources to prepare for this climate-induced crisis.

However, the government continued its policy of subsidizing the fossil fuel industry with an annual amount of $8 billion. Even though Australia records the highest solar radiation per square meter in the world, solar power subsidies were withdrawn and hence, photovoltaic energy generation stands at a very low level in the continent. Figure 11.2 shows Australians protesting government inaction against climate change by forming various public rallies.

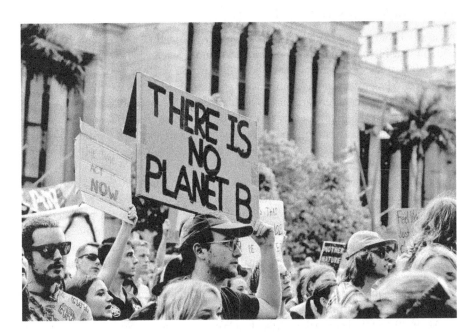

**FIGURE 11.2** Australians protesting government inaction against climate change.

An academic review of 57 scientific papers, which have been published since 2013, was executed by Sciencebrief.org, an online platform for research on global climate change. The review indicated an association between climate change and enhanced bushfire seasons across the globe (Betts et al. 2020). Richard Betts, one of the co-authors of the review and Head of Climate Impacts Research at Britain's Met Office Hadley Centre, said in a conference held in London that across the globe, fire seasons across the globe have expanded by about 25% of Earth's vegetated surface. He further stated that climate changes had warmed Australia's land area in excess of the average global temperature rise of 1 degree Celsius (Al Jazeera 2020).

In 2019, Australia recorded not only the driest year ever but also the hottest. The country witnessed average temperatures of 1.52 degrees Celsius above the 1961–1990 average. New South Wales recorded temperatures of 1.95 degrees Celsius above average (Readfearn 2020). The year was marked with several deficiencies in precipitation levels, about 40% less than average based on 1990 records, and hydrological drought that exacerbated high fire dangers in the biodiversity-rich regions of southern and eastern Australia.

By spring 2019, over 95% of the country had recorded the highest fire danger data in terms of Forest Fire Danger Index (FFDI) that was much above the regular average (Australian Government Bureau of Meteorology 2019). With FFDI values of 100 or above in New South Wales, the season marked the onset of humongous bushfires on September 6, 2019, in the eastern part of the country. The temperatures recorded in 2019 until the spring season varied between "very much above average" and "highest on record over large areas."

## 11.3  IMPACT ON WILDLIFE

Australia is one of the world's most biodiversity-rich nations and ranks 17th worldwide in housing a megadiversity of flora and fauna. Scientists have estimated that a mere 30% of species available have been discovered yet over 600,000 species are predicted to be present in the continent (Australian Academy of Science 2020).

The bushfires scourged large areas of land rich in biodiversity and destroyed the habitats of several animal species native to the continent. Analysts suggest that the 2019–2020 bushfire season has led to the death of about one billion animals, including domestic pets and livestock, and has led to a drastic scarcity of food and shelter for millions of others.

The fires, considered to be an "ecological disaster," have led to the destruction of native flora and fauna. As per a statement issued by the World Wildlife Fund (WWF), areas like Western Australia's Stirling Ranges, Australian Alps, Gondwana Rainforest, and the Blue Mountains have witnessed catastrophic damage. The animals comprising the casualty toll include species of Koala, black glossy cockatoo, western ground parrot, eastern bristlebird, the Kangaroo Island dunnart, Blue Mountains water skink, long-footed potoroo, brush-tailed rock wallaby, pygmy possums, gliders, mountain lizards, pear-shaped frog, and leaf-tailed gecko, in addition to numerous other animal and bird species.

About 30% of the koala habitat (Eucalypt Woodlands in New South Wales) has been estimated to be lost because of the natural disaster, as per Sally Box, Threatened Species

**FIGURE 11.3**   Impact of fires on koala habitat.

Commissioner, and Sussan Ley, Environment Minister of Australia. Approximately 25,000 koalas are estimated to have died in the blazing fire (Al Jazeera 2019). Figure 11.3 shows a koala crying for help during a bushfire as its habitat gets burnt.

Moreover, according to Storm Standford, Flying Fox coordinator at WIRES Australian Wildlife Rescue Organisation, 50% of infant grey-headed flying foxes born in 2019 died in the devastating bushfire season (Shelton 2020).

A mission named "Operation Rock Wallaby" was carried out to safeguard the rock wallabies. The rock wallabies survived the fires but were faced with a dearth of food and habitat in the aftermath of the crisis. The national park staff employed helicopters to airdrop thousands of sweet potatoes and carrots to ensure the continuation of the endangered species of brush-tailed rock wallabies in the surrounding areas of New South Wales (Sonali 2020).

It is estimated that around 40% of the bushfires blazed over conservation zones and national parks, areas that are under government protection (Munroe and Taylor 2020). The country contains a forest area equivalent to the fourth-largest forest area globally and is significant for biodiversity.

## 11.4   IMPACT ON THE ECONOMY

The bushfire's toll on the Australian economy is inconceivable in magnitude. The fires have had direct and indirect effects on the economy, with both immediate and long-term impacts projected. The disaster cut through the country's coasts during the summer months, affecting the earning potential of Australia's tourism industry, the dairy industry, and the agricultural sector.

Analysts have estimated that the bushfires have cost the Australian economy $3.5 billion.

Australia has had to account for the immediate costs associated with fire-fighting, relocating residents, medical treatments owing to burns and respiratory illnesses, work days lost, shaken consumer confidence, asset destruction costs related to businesses, farms, and housing, along with the replacement and rebuilding costs. The long-term costs include treating post-traumatic stress disorders (PTSD) and addressing the damage to ecosystems, including water contamination and the extinction and deaths of a variety of animal and bird species, especially those that are necessary for agricultural production, like bees. Long-term costs also include factors that directly influence the tourism industry and permanently decrease tourism, reduced foreign investment in the economy, and an increase in the cost of living and insurance costs in the aftermath of the crisis. The bushfires have shaken consumer confidence, impacting the overall demand in the economy. According to the Melbourne Institute and Westpac Bank survey, in January, consumer sentiment data, a chief indicator in Australia that predicts the real spending data of consumers, was recorded to be 6.2% lower than the year before (Westpac 2020). Consumer sentiment data is deemed to be a chief indicator in the country that predicts the real spending data of consumers. The survey of consumer confidence conducted by ANZ-Roy Morgan recorded a level lower than the levels recorded in the last four years (Morgan 2020). Katrina Ell, Moody's economist, revealed that the impact of bushfires on consumer confidence and economic damage owing to macroeconomic spillovers increased the chances of rate cuts by Australian authorities. Consumer prices have shot up because of the damages to fresh produce, and areas like Kangaroo Island have seen failing farms and 90% of the timber plantations impacted by the fire. The disaster has destroyed one-third of grapevines, which had a drastic impact on the wine industry in the Adelaide Hills. The bushfires destroyed the paper mill in Eden, a ski-resort in Mt. Selwyn, plantations of softwood and pine in Tumut, and Batlow's apple orchards. Even with the financial aid packages provided by the government, these regions may need significant recovery time, such as five years anticipated recovery time for Batlow's apple orchards and twenty years for Tumut's pine plantations.

Analysts suggest that the bushfires could cost the $1.33 trillion Australian economy more than $4.4 billion, which was the cost of the 2009 bushfires in the country. The bushfires have a projected expense that could erase 0.25%–1% from the national gross domestic product (GDP) in the December and March quarters. This is compared to 0.4% recorded in the latest quarter, a change that could ultimately lower Australia's growth projections (Fensom 2020). The daily economic activity in Sydney is about A$1.2 billion, but the onset of the bushfire season has caused daily disruptions between A$12 million and A$50 million in regional economic activity (McDonald 2019).

In response to the crisis, the Prime Minister of Australia, Scott Morrison, has established a National Bushfire Recovery Agency to engage in the activities of rebuilding communities. Australian administration has allocated $1.4 billion in the form of federal aid for the rebuilding of health and education infrastructure (Regan 2020), with an additional A$11 million pledged for the National Aerial Firefighting Centre. Figure 11.4 shows an Erickson Air Crane helicopter dropping loads of water over a bushfire in Bundoora, Australia.

**FIGURE 11.4**    Erickson air crane helicopter.

Future plans include an additional allocation of A$500 million for the fiscal year ending June 2020 and approximately A$1 billion for 2020–2021 (Cole 2020). At this rate, the Climate Council estimates that the cost of the bushfires devastation could reach A$19 billion by 2030 and about A$4 trillion by 2100 as a result of reduced labor and agricultural productivity in the continent (Bloomberg 2020).

## 11.5   IMPACT ON THE INSURANCE INDUSTRY

The insurance industry in Australia has been no stranger to the disasters that have impacted the country over the decades. In terms of losses insured, the most exorbitant bushfire seasons in Australia have been the year 1983 with an expense of about $1.76 billion and the year 1967 with an expense of $2.16 billion (Demetriadi 2020). However, the losses from 2019 seem striking even in comparison to these other catastrophic events, and analysts have been unable to predict whether the Australian insurance industry can survive the unprecedented hit.

Insurance Australia Group Ltd (IAG) has quoted their approximate claim costs, arising from natural disasters, at $279 million for the half fiscal year 2020. IAG, the largest insurer, stated that it received more than 2,800 claims associated with bushfires, adding up to approximately $111 million of the aggregate estimated insurance expense arising from natural disasters (Reuters 2020). Having received over 2,600 claims as of January 9, 2020, Suncorp Group Ltd has estimated losses ranging from A$315 million to A$345 million (Ullah et al. 2020).

The Insurance Council of Australia (ICA) released approximations that put the total claims estimate to 20,000, totaling to a sum of $1.65 billion of value loss owing to bushfires of 2019.

Regarding the capacity of insurers to absorb the cost, Michael Vine, Director of the S&P Global Ratings team dealing in financial services ratings, stated that the Australian insurance market was accustomed to such natural perils, which permitted insurers to make necessary allowances in their forecasts and estimates. This view is also supported by A.M. Best, a credit rating agency in the United States. Insurers' capital positions and high reserve adequacy ratios, combined with previous strong underwriting performance, lead to general conclusions that even if the predictions and stop-loss mechanisms failed, their aggregate reinsurance covers could help the industry in meeting the claims and absorb the losses (Demetriadi 2020).

As per Moody's, the bushfire season can have a major impact on companies like IAG Ltd, Suncorp Group, QBE, Allianz SE, and Zurich Insurance (Martin 2020b). However, the application of reinsurance capital is allowing insurers to dilute their concentration risks. IAG Ltd, for example, increased its gross reinsurance buffer to about $6.3 billion for the year 2020 (Reuters 2020). Moreover, insurers have implemented measures to reduce the devastation potential of the mega bushfires, including financial relief measures like premium waivers, charitable donations, mental health support systems, and deploying helicopters to aid rural firefighters. IAG's National Roads and Motorists' Association (NRMA) Insurance has assisted the Rural FireService (RFS) in New South Wales by using the NRMA helicopter to deploy non-toxic and biodegradable fire retardant in the communities of Batemans Bay and Moruya to protect properties and assets from catching fire. The helicopter also contributed to water-bombing missions in areas like Gospers Mountain fire, Cessnock, New South Wales Southern Coast, and Turramurra (Martin 2020a).

Additional responses from insurers to reduce claim liability come in the form of embargoes. The industry players undertake this practice to prevent the sale of new policies in areas prone to disasters and to deny claim liability arising out of new policies in the areas affected by disasters.

Such a measure prevents people from panic-buying insurance policies to safeguard their financial assets directly before a catastrophe. Likewise, IAG Ltd. and Suncorp Group embargoed 67 areas in parts of New South Wales and Victoria that were affected by the bushfires. The embargoes were lifted on an individual basis only when authorities could control the fires. Non-renewal clauses were not imposed on existing policies. However, companies like QBE and Westpac did not use embargoes. Instead, Westpac imposed a waiting period of seven days before which new policy owners could not raise their claims against insured properties (Butler 2020).

The bushfires have exposed properties in every 4.5 square kilometer compared to 1 square kilometer during previous disasters. Also, certain limitations, like distance covered by disasters and disaster durations, restrict the scope of reinsurance contracts. Such unprecedented catastrophes can push up premiums for customers owing to higher reinsurance costs.

The biggest fear that compromises the positive effects of financial aid packages and claims acceptances by insurers is the practice of underinsurance often adopted by insurance policyholders to reduce their premium liabilities. However, this practice ultimately leads to insufficient funds available in the hands of the aggrieved claimants to be able to rebuild their properties or restore their assets.

Another issue concerning the insurance industry of Australia is the unavailability of standard terms, definitions, and clauses on "fire" in insurance policies that make the nation's citizens more skeptical of trusting insurance companies. CHOICE, a member-funded and independent consumer advocacy group operating in Australia, acknowledged in their surveys that about 35,000 Australians did call for clear fire definitions and the company identified unclear "fire" definitions and complex terminologies in 70% of policies with some serious issues in policies accounting to about 25% of the total volume. The group has called for government intervention in establishing standardized definitions concerning fires so that a possible collaboration between the two macroenvironment forces, the government, and the insurance industry, can better safeguard the interests of the general public from known perils and natural calamities (Kirkland 2020).

Even though the Australian insurance industry bears the ability to cope with the disaster, the current picture all the more provides evidence of the fact that the environmental risk poses an important bearing on future predictions that need to be accounted for by the players to reduce their aggregate exposures. As per Frank Mirenzi, Senior credit officer and Vice-President at Moody's, the property and casualty insurance industry is prone to the devastating effects of climate change and its impact in the form of severity or frequency of catastrophic events. Figure 11.5 shows property losses to Australian commercial and residential market from previous Australian bushfires.

In the urge to mitigate risks arising from environmental hazards and build environment adaptability, insurers are employing tools, like scenario analyses, to ascertain the full-scale impact of climate change (Martin 2020).

## 11.6 IMPACT ON LIVES, LAND, AND POLLUTION

The most devastating bushfire season ever recorded in the country and bigger than California came to a close by the end of March 2020. The fires blazed through more than 10.3 million hectares of land, equivalent to the size of South Korea (Scarr et al. 2020). New South Wales was significantly impacted with more than 6% of the state's land burned, amounting to an area of about 5.5 million hectares. The state recorded a four-time increase in fire alarms in 2019–2020 versus over the past two decades, with about 9,360 alarms in 10 days since the start of 2020 (Munroe and Taylor 2020). As seen in Figure 11.6, there was a steep rise in fire alarms received by NSW in 2019 when compared to alarms received in previous years.

Figures published from the respective emergency fire services state that the three states of Victoria, Queensland, and New South Wales accounted for over 84% of the bushfires. The fires destroyed national parks and last-measure resorts, such as military planes and warships, had to be employed to evacuate Australia's citizens.

The fires demolished about 2,500 homes, displaced 90,000 people, and claimed the lives of over 33 people, including the lives of fire service volunteers and firefighters. The fires increased air pollution levels in Canberra, Sydney, and Melbourne to alarming levels, making them record worse air quality index ratings on a global level. As per readings from the office of Global Modeling and Assimilation, NASA, the fire's smoke, representing the release of organic carbon, spread to an area of about 17 million square kilometers (Hernandez et al. 2020). Such organic carbon consists

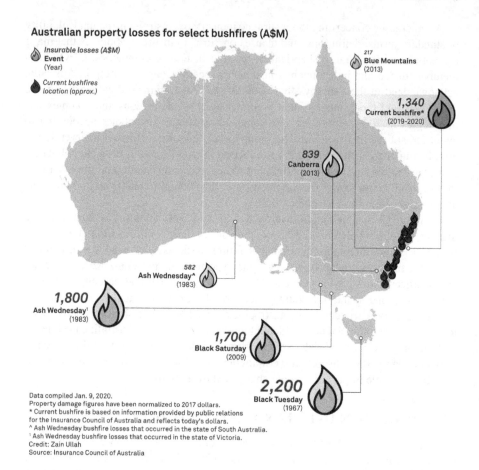

**Australian property losses for select bushfires (A$M)**

FIGURE 11.5 Property losses from Australian bushfires. (Rendering credit: Zain Ullah, S&P Global Market Intelligence.)

of particles like Particulate Matter 2.5, which is linked to causing cancer, heart strokes, and other diseases. PM 2.5 exposure rates recorded in Goulburn, NSW, stood at 2,000 micrograms, far exceeding the "hazardous level" identified at 250 micrograms on air quality indication classifications of the U.S. Environmental Protection Agency. As seen in Figure 11.7, bushfire smoke covered the Sydney Opera House and a large cruise liner on December 10, 2019, reducing visibility to alarmingly low levels. The bushfire season has contributed to emissions of carbon dioxide of over 400 megatons as per the estimates of the European Union's Copernicus monitoring program (UNEP 2020). Also, the United Nations World Meteorological Organization has expressed concerns that the fire deposited soot on glaciers of New Zealand, accelerating their liquifying rates.

Prime Minister Scott Morrison has opened an enquiry into the bushfires and various reasons that led to the crisis. For the purpose, a Royal Commission has been established for six months that would inspect bushfire preparedness of the economy

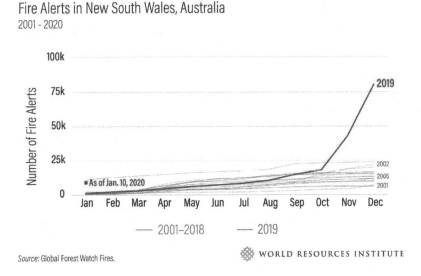

**FIGURE 11.6**   Fire alarms received by NSW. (Rendering credit: Global Forest Watch.)

in the immediate future and need for any law modifications to shed light on the role of various departments and authorities in the events to come (Packham 2020).

The bushfire season is extended by dust storms, extended periods of drought, heatwaves, and low precipitation levels, and burning the flora releases $CO_2$ in significant concentrations. This develops into a positive feedback loop in which high carbon emissions lead to more severe changes in weather patterns that ultimately intensify bushfires.

**FIGURE 11.7**   Sydney on December 10, 2019.

## 11.7 TACKLING THE CRISIS: EXISTING AND IMMINENT

This hour calls for innovative scientific solutions that can cope with unpredictable future scenarios, along with active collaborations between public and private sectors and multinational organizations, to provide the global economies a fighting chance in tackling the impending climate crisis.

One key method to reducing losses lies in the use of high-resolution loss simulation models. Munich RE, for instance, has employed a Software as a Service (SaaS) Natural Hazards Assessment Network (NATHAN) that enables the assessment of location-specific risks as well as total risk envelopes and the enhancement of claims management. NATHAN allows risk modeling, consists of individual risk score ratings, and is based on a collective database lasting decades of the world's disaster data. Additionally, this risk assessment tool provides a baseline to risk managers and insurance underwriters for risk and trend evaluation and benchmarking. Furthermore, catastrophe databases like NatCatSERVICE allow people to customize needs and mine data from records that provide decades of information to ensure more accurate evaluation (Munich RE 2020a).

Australia employs the use of satellites to identify fires; the Sentinel-2 satellite from Europe, for instance, provides a decent check on vegetation through its infrared channels of a short wavelength. Scientists in Australia are also in the course of developing a satellite to aid authorities in better identification of inception points of bushfires in the immediate future. The satellite would assist in evaluating the moisture content of forests in the country and the "fuel load," that is, the quantity of flammable substances present around a fire. The feat will be achieved through infrared detectors installed on the spacecraft tuned to the country's eucalypt shrubs and trees. It has been predicted that the detectors could trace reflected light into thin bands that characterize eucalypt species properties that would allow them to detect minute changes, like water content, number of leaves, and lignin content, in spectral signatures of the available flora (Amos 2020).

As the world acknowledges the significance of climate change and its influential impact on risk assessments across the globe, it has become imperative to prepare for existing and emerging hazards arising from natural variability in climate conditions. Examination and analysis of perils that are not yet provided for in the risk models are required to make precise assessments of a hazard's loss potential and to gather preliminary information about events at their nascent stages. All concerns imply the inclusion of environmental or climate-oriented risks in model simulations to ensure the existence of flora and fauna and our own in the decades to come.

## REFERENCES

Al Jazeera. 2019. "Thousands of koalas feared dead in Australia wildfires." https://www.aljazeera.com/news/2019/12/thousands-koalas-feared-dead-australia-wild-fires-191228085136930.html.

Al Jazeera. 2020. "Melbourne chokes amid warning gigantic fires could become routine." https://www.aljazeera.com/news/2020/01/melbourne-chokes-warning-massive-fires-routine-200114033251588.html.

Amos, Jonathan. 2020. "Bushfires: Australian satellite would be 'tuned' to eucalypt vegetation." *BBC*. https://www.bbc.com/news/science-environment-51727231.

Australian Academy of Science. 2020. "The Australian bushfires—why they are unprecedented." https://www.science.org.au/news-and-events/news-and-media-releases/australian-bushfires-why-they-are-unprecedented.

Australian Government Bureau of Meteorology. 2019. "Special Climate Statement 72—dangerous bushfire weather in spring 2019." http://www.bom.gov.au/climate/current/statements/scs72.pdf.

Betts, Richard, Matthew W. Jones, Adam Smith, Josep G. Canadell, I. Colin Prentice, and Corinne Le Quéré. 2020. "Climate change increases the risk of wildfires." *ScienceBrief*. https://sciencebrief.org/briefs/wildfires.

Bloomberg. 2020. "Australia's wildfire crisis key numbers behind the disaster." *The Economic Times*. https://economictimes.indiatimes.com/news/international/world-news/australias-wildfire-crisis-key-numbers-behind-the-disaster/articleshow/73093363.cms?utm_source=contentofinterest&utm_medium=text&utm_campaign=cppst.

Butler, Ben. 2020. "Economic impact of Australia's bushfires set to exceed $4.4bn cost of Black Saturday." *The Guardian*. https://www.theguardian.com/australia-news/2020/jan/08/economic-impact-of-australias-bushfires-set-to-exceed-44bn-cost-of-black-saturday.

Butler, Ben. 2020. "Suncorp and IAG temporarily stop selling insurance in fire-affected areas of Victoria and NSW." *The Guardian*. https://www.theguardian.com/business/2020/jan/14/suncorp-and-iag-temporarily-stop-selling-insurance-in-fire-affected-areas-of-victoria-and-nsw.

Cole, Wayne. 2020. "Australia commits extra A$2 billion for bushfire rebuilding." *Reuters*. https://in.reuters.com/article/us-australia-bushfires-assistance/australia-commits-extra-a2-billion-for-bushfire-rebuilding-idINKBN1Z509K.

Demetriadi, Alexi. 2020. "Can the insurance industry sustain bushfire losses?" *Insurance Business Australia*. https://www.insurancebusinessmag.com/au/news/breaking-news/can-the-insurance-industry-sustain-bushfire-losses-211716.aspx.

Fensom, Anthony. 2020. "Up in smoke: Australia's bushfires darken economic outlook." *The Diplomat*. https://thediplomat.com/2020/01/up-in-smoke-australias-bushfires-darken-economic-outlook/.

Global Forest Watch. 2020. "4 things to know about Australia's wildfires and their impacts on forests." *World Resources Institute*. https://www.wri.org/blog/2020/01/4-things-know-about-australia-s-wildfires-and-their-impacts-forests.

Hernandez, Marco, Manas Sharma, and Simon Scarr. 2020. "Graphic: Australia's bushfires make massive clouds of pollution." *Reuters*. https://in.reuters.com/article/australia-bushfires-smoke/graphic-australias-bushfires-make-massive-clouds-of-pollution-idINKBN1Z915Q.

Kirkland, Alan. 2020. "Alan Kirkland: 'When is a fire not a fire?'." https://www.choice.com.au/money/insurance/home-and-contents/articles/choice-ceo-alan-kirkland-op-ed-bushfire-definitions.

Martin, Mina. 2020a. "IAG's NRMA supports bushfire-fighting efforts in NSW." *Insurance Business Australia*. https://www.insurancebusinessmag.com/au/news/breaking-news/iags-nrma-supports-bushfirefighting-efforts-in-nsw-196017.aspx.

Martin, Mina. 2020b. "Moody's: Escalating bushfire insurance losses will remain manageable for Australian P&C insurers." *Insurance Business Australia*. https://www.insurancebusinessmag.com/au/news/breaking-news/moodys-escalating-bushfire-insurance-losses-will-remain-manageable-for-australian-pandc-insurers-196217.aspx.

Mcdonald, Tim. 2019. "Australia fires: The huge economic cost of Australia's bushfires." *BBC*. https://www.bbc.com/news/business-50862349.

Munich RE. 2020a. "NATHAN for location risk intelligence." https://www.munichre.com/en/solutions/reinsurance-property-casualty/nathan.html.

Munich RE. 2020b. "Wildfires." https://www.munichre.com/en/risks/natural-disasters-losses-are-trending-upwards/wildfires-as-the-climate-changes-so-do-the-risks.html.

Munroe, Thailynn, and Rod Taylor. 2020. "4 things to know about Australia's wildfires and their impacts on forests." *World Resources Institute.* https://www.wri.org/blog/2020/01/4-things-know-about-australia-s-wildfires-and-their-impacts-forests.

National Geographic Kids. 2020. "What is a bushfire?" https://www.natgeokids.com/uk/discover/geography/physical-geography/what-is-a-bushfire/.

Packham, Colin. 2020. "Australia begins wide-ranging enquiry into deadly bushfires." *Reuters.* https://www.reuters.com/article/us-australia-bushfires/australia-begins-wide-ranging-enquiry-into-deadly-bushfires-idUSKBN23103R.

Readfearn, Graham. 2020. "Explainer: what are the underlying causes of Australia's shocking bushfire season?" *Guardian.* https://www.theguardian.com/environment/2020/jan/13/explainer-what-are-the-underlying-causes-of-australias-shocking-bushfire-season#maincontent.

Regan, Helen. 2020. "Australia's bushfire royal commission to focus on preparing for future emergencies, but not climate change policy." *CNN.* https://edition.cnn.com/2020/02/20/australia/australia-fires-royal-commission-intl-hnk/index.html.

Reuters. 2020. "Australia's IAG expects $280m claims for natural disasters." *Al Jazeera.* https://www.aljazeera.com/ajimpact/australia-iag-expects-280m-claims-natural-disasters-200103034527976.html.

Roy Morgan. 2020. "ANZ-Roy Morgan Consumer Confidence plunges to lowest historical record at 65.3." http://www.roymorgan.com/findings/8345-anz-roy-morgan-consumer-confidence-march-31-2020-202003302354.

Scarr, Simon, Manas Sharma, and Marco Hernandez. 2020. "Sizing up Australia's bushfires." *Reuters.* https://in.reuters.com/article/uk-australia-bushfires-scale-graphic/sizing-up-australias-bushfires-idINKBN1Z81FN.

Shelton, Tracey. 2020. "'A shape in the ash': Bushfires destroy Australian wildlife." *Al Jazeera.* https://www.aljazeera.com/news/2020/01/shape-ash-bushfires-destroy-australian-wildlife-200103005254387.html?utm_source=website&utm_medium=article_page&utm_campaign=read_more_links.

Sonali, Paul. 2020. "Koalas, wallabies endangered by Australia bushfires 'ecological disaster'." *Reuters.* https://in.reuters.com/article/australia-bushfires-wildlife/koalas-wallabies-endangered-by-australia-bushfires-ecological-disaster-idINKBN1ZC0DD.

Spence, Bill. 2020. "Australian government is burning our children's future." *Al Jazeera.* https://www.aljazeera.com/indepth/opinion/australian-government-burning-children-future-200213100102590.html.

Stevens, Kylie. 2019. "A nation in flames: The shocking toll of summer bushfires in Australia." *Daily Mail Australia.* https://www.dailymail.co.uk/news/article-7838193/How-current-bushfire-crisis-compares-biggest-blazes-Australias-history.html.

Ullah, Zain, Jason Woleben, and Hailey Ross. 2020. "Australia's largest insurers brace for greater bushfire severity as claims grow." *S&P Global Market Intelligence.* https://www.spglobal.com/marketintelligence/en/news-insights/latest-news-headlines/56476911.

Understand Insurance. 2020. "Bushfires." http://understandinsurance.com.au/types-of-disasters/bushfires.

UNEP. 2020. "Ten impacts of the Australian bushfires." https://www.unenvironment.org/news-and-stories/story/ten-impacts-australian-bushfires.

Westpac. 2020. "Consumer Sentiment falls further." https://www.westpac.com.au/content/dam/public/wbc/documents/pdf/aw/economics-research/er20200122BullConsumerSentiment.pdf.

# Section IV

## Green Engineering

# 12 Concrete Containing Waste Materials Considering Environment Sustainability

*Damyanti Badagha, Niragi Dave,*
*Pratik Chauhan, and Kaushal Parikh*

## CONTENTS

## 12.1 INTRODUCTION

Rapid urbanization and population growth have increased the rate of infrastructure development all over the world, making the construction industry one of the fastest developing sectors globally (Govindan et al. 2016). The concrete industry faces the challenge of attaining durable concrete with adequate strength in a way that is both economical and environmentally sustainable (Swamy 2008).

Iron is the most common metal used worldwide. Iron ore, coal, limestone, and recycled steel are the main raw materials required to produce iron. As per Sadek (2014), 270–320 Mt iron slag was produced in 2012, with increasing demand every year. Any improvement in energy recovery, recycling of the by-product waste, and recycling during the process of iron manufacture will have positive global implications, both economically and environmentally (Zhao et al. 2015).

The iron and steel-making process generates significant amounts of waste materials as slag (Kan et al. 2015). Due to landfill restrictions and other environmental considerations, recycling these by-products has now become a major problem for this industry, which needs to preserve natural resources and conserve energy with the use of material with suitable properties and characteristics (Motz and Geiseler 2001).

According to Dimitrova (1995) and Ozturk and Gultekin (2015), slag can be considered a resource, depending on its properties, since it can be used in another part of the industry, which helps to reduce environmental pollution, energy use, and production costs. Growing global concerns about the environmental, social, and economic impacts of carbon dioxide ($CO_2$) and other greenhouse gas (GHG) emissions have motivated the development of various tools and strategies for mitigating $CO_2$ emissions (Proske et al.2013; Fais et al. 2016). Most of the previous research in this area only considers energy and $CO_2$ emission reduction strategies. Researchers have expanded key decision-making criteria to include economics and profitability (Zainuddin and Wan 2017).

The solution to this environmental challenge is to incorporate waste material into concrete to minimize using cement, leading to less cement manufacture, which in turn brings about a reduction in $CO_2$ emissions. High performance concrete (HPC) that uses only cement as a binder requires a high paste volume, which often leads to excessive shrinkage, a large amount of hydration heat, and increased cost. Ground granulated blast furnace slag (GGBFS) used as a partial replacement of cement has been proven to overcome these problems and improve the concrete's durability. The HPC that contains GGBFS is denser and more impermeable than conventional concrete (Rajamane et al. 2003). Hannant (1967) and Nasser and Neville (1965) observed the long-term mechanical behavior of HPC regarding structural serviceability, such as tension losses in strength, excessive deflection, and cracking with temperature changes.

Several researchers looked at the influence of using GGBFS on the elastic modulus of concrete, and it was concluded that elastic modulus, as compared to compressive strength, is less important to augmented GGBFS substitution due to the decrease in porosity by the further formation of calcium silicate hydrate as a result of pozzolanic reaction, especially at the later ages such as 180 days. However, a consensus has not been reached about the influence of GGBFS on the static elastic modulus of concrete. The impact of GGBFS substitution up to 60% with 20% increment on the static elastic modulus of high strength concretes at 28 days of age was studied and the static elastic moduli of concrete with 0%, 20%, 40%, and 60% GGBFS were found to be 44.58, 45.02, 44.84, and 43.09 MPa, respectively. Concretes containing up to 60% GGBFS had equal or slightly higher static elastic modulus values when compared with concrete containing only Portland cement (PC). A slight difference was perceived between the static modulus of PC concrete and that of the concrete produced with GGBFS originating in South Africa. Also, no meaningful difference was found regarding elasticity moduli between concrete produced with Japanese GGBFS-PC (PC) and only PC. In contrast, lower elasticity modulus values were measured for the concretes with the United Kingdom GGBFS-PC compared with concretes with only PC; this is valid only at young ages and high volumes of GGBFS use. Swamy and Bouikni (1990) reported little change between the elasticity modulus of concrete with and without GGBFS.

## 12.2 MATERIALS AND METHODS

### 12.2.1 MATERIALS

In hydraulic cement concrete, the binder is composed of a mixture of hydraulic cement and water. This research study used 53-grade OPC (ordinary Portland cement), confirming IS 12269:2013. The OPC cement, was tested to initially ensure the quality during the experimental period, along with at the end of each work phase. The cement properties of consistency, initial setting time, final setting time, soundness, and fineness were found to be 30.25%, 125 minutes, 240 minutes, 1 mm expansion, and 1-gram retention on a 90-micron sieve, respectively. The measure of the compressive strength at 3 days, 7 days, and 28 days was found to be 29.5 N/mm$^2$, 39.70 N/mm$^2$, and 56.70 N/mm$^2$, respectively.

Locally available crushed stones and river sand were used as fine aggregate and coarse aggregate, confirming IS 383:1970. Table 12.1 shows the properties of aggregates.

This research used waste from the local steel industry in Surat, India. The original form of the waste was granular. With the help of a ball mill, slag was converted into a very fine powder form. Table 12.2 presents the slag's physical and chemical analysis.

As per the data in Table 12.2, the waste powder's physical and chemical characteristics are similar to cement's physical and chemical characteristics. The compressive strength of mortar mixes was found by replacing 10%–90% of the cement by waste, with an interval of 10% (shown in Figure 12.1). It was found that 50% cement replacement gives the maximum increased strength. The mixed design was completed for conventional concrete and concrete containing 50% cement replacement by industrial waste.

### TABLE 12.1
### Properties of Aggregates

| Test for Aggregate | Sand | 10 mm Aggregate | 20 mm Aggregate |
|---|---|---|---|
| Specific gravity | 2.55 | 2.8 | 2.83 |
| Water absorption | 1.21 | 1.21 | 1.21 |
| Bulk density (kg/m$^3$) | 1565 | 1392 | 1437 |
| Fineness modulus | 2.54 | 2.93 | 4.66 |
| Elongation index | — | 14.08 | 14.23 |
| Flakiness index | — | 13.83 | 13.27 |
| Silt content (%) | 1.95 | — | — |
| Grading zone | Zone III | 10 mm aggregate | 20 mm aggregate |
| Crushing value (%) | — | 15.12 | 17.98 |
| Impact value (%) | — | 11.22 | 13.39 |

## TABLE 12.2
## Physical and Chemical Characteristics of Waste Powder

| Sr. No | Characteristic | Requirement as per BS:6699 | Test Result |
|---|---|---|---|
| 1 | Fineness (m²/kg) | 275 (Min.) | 420.00 |
| 2 | Residue by wet sieve (%) | — | 2.00 |
| 3 | Initial setting time (min) | 30 min (Min.) | 195 |
| 4 | Insoluble residue (%) | 1.5 (Max.) | 0.28 |
| 5 | Magnesia content (%) | 14.0 (Max,) | 8.06 |
| 6 | Sulfide sulfur (%) | 2.00 (Max.) | 0.53 |
| 7 | Sulfate content (%) | 2.50 (Max.) | 0.24 |
| 8 | Loss on ignition (%) | 3.00 (Max.) | 0.29 |
| 9 | Manganese content (%) | 2.00 (Max.) | 0.23 |
| 10 | Chloride content (%) | 0.1 (Max.) | 0.001 |
| 11 | Moisture content (%) | 1.00 (Max.) | 0.03 |
| 12 | Glass content (%) | 67 (Min.) | 94.00 |
| **Chemical Moduli** | | | |
| 13 | $CaO + MgO + SiO_2$ | 66.66 (Min.) | 82.00 |
| | $CaO + MgO/SiO_2$ | 1.0 (Min.) | 1.3 |
| | $CaO/SiO_2$ | 1.40 (Max.) | 1.06 |

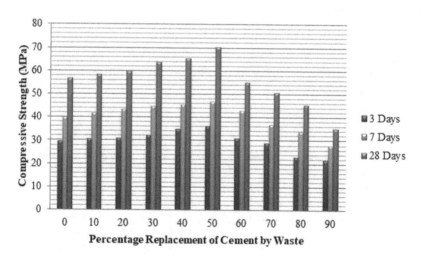

FIGURE 12.1   Compressive strength of mortar with waste as cement replacement.

## TABLE 12.3
## Slag Activity Index for Percentage Variation of Waste in Waste-Cement Mortar

| Slag Activity Index | Percentage of Waste in Waste-Cement Mortar | | | | | | | | |
|---|---|---|---|---|---|---|---|---|---|
| | 10 | 20 | 30 | 40 | 50 | 60 | 70 | 80 | 90 |
| 7 Days | 105 | 109 | 113 | 114 | 118 | 108 | 93 | 85 | 70 |
| 28 Days | 103 | 105 | 112 | 115 | 124 | 97 | 89 | 80 | 62 |

As per ASTM C989, the slag activity index is the percentage ratio of the average compressive strength of slag-cement (50%–50%) mortar cubes to the average compressive strength of reference cement mortar. Table 12.3 shows the slag activity index values for different percentages of waste in waste-cement mortar.

$$\textit{Slag activity index} = \frac{SP}{P} \times 100 \qquad (12.1)$$

$SP$ = average compressive strength of slag-cement mortar cubes in MPa, and
$P$ = average compressive strength of cement mortar cubes in MPa.

The slag activity index for the 50% replacement of cement by waste is 118 for 7 days and 124 for 28 days, indicating the grade for waste material is Grade 120.

### 12.2.2 CONCRETE COMPOSITION

The composition of concrete containing waste powder as 50% cement replacement has been established after several trial mixes. Table 12.4 contains the composition of concrete containing waste powder and conventional concrete for three different grades of concrete: M20, M40, and M60. The specimens were casted and cured in tap water for 28 days.

## TABLE 12.4
## Mix Proportions

| Materials | CC20 | CW20 | CC40 | CW40 | CC60 | CW60 |
|---|---|---|---|---|---|---|
| OPC 53 grade (kg) | 300 | 150 | 465 | 235 | 510 | 265 |
| Waste powder (kg) | — | 150 | — | 235 | — | 265 |
| 20 mm aggregate (kg) | 1029 | 669 | 728 | 740 | 664 | 696 |
| 10 mm aggregate (kg) | 400 | 455 | 423 | 428 | 441 | 464 |
| Sand (kg) | 696 | 906 | 531 | 680 | 687 | 648 |
| Admixture (kg) | — | 2.7 | — | 4.05 | 3.12 | 3.92 |
| Water (kg) | 150 | 163 | 186 | 162 | 164 | 159 |
| W/C ratio | 0.5 | 0.54 | 0.4 | 0.36 | 0.32 | 0.30 |

CC-conventional concrete, CW-concrete with 50% waste.

## 12.2.3  Testing

The compression test was performed on cube specimens with edges of 150 mm, as per IS 516:1959.

## 12.2.4  Modulus of Elasticity (Es)

The static modulus of elasticity tests were carried out on cylindrical specimens, as per the guidelines of ASTM C 469-02. As shown in Figure 12.2, dial gauges with the least count of 0.01 mm were attached. The load was gradually applied. The modulus of elasticity was determined as the slope of the secant, drawn from the origin to a point corresponding to 40% of the cylinder's compressive strength of concrete on the stress–strain curve. The experimental value of the modulus of elasticity was compared with the calculated values, as per IS 456-2000, ACI-209, and BS-8110.

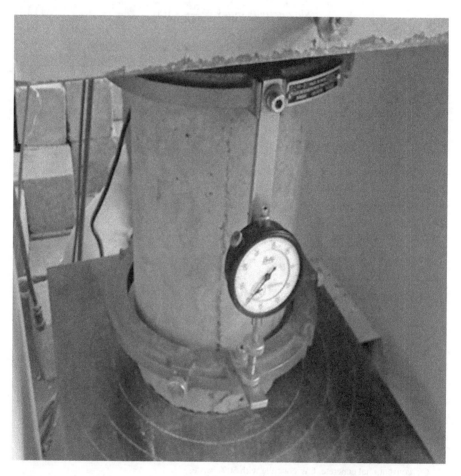

**FIGURE 12.2**  Specimen arrangement for determining modulus of elasticity.

The following equations were used to calculate modulus of elasticity:

$$\text{From IS 456:2000, } E_s = 5000\sqrt{f_c} \tag{12.2}$$

$$\text{From ACI-209, } E_s = 0.043 \times \sqrt{\frac{\rho^3 f_c t}{4 + 0.85t}} \tag{12.3}$$

$$\text{From BS 8110:1997, } E_s = \left(20 + 0.25 f_c\right) \times \left(0.4 + 0.6\frac{f_c'}{f_c}\right) \tag{12.4}$$

## 12.3 RESULTS AND DISCUSSION

The compressive strength of concrete containing 50% waste is more than that of conventional concrete, as shown in Figure 12.3. The compressive strength for concrete containing waste is 8.83%, 83.55%, and 4.96% higher than that of conventional concrete for M20, M40, and M60 grade concrete, respectively.

Table 12.5 and Figure 12.4 show comparisons of experimental modulus of elasticity (MPa) with the values obtained from different standards. Table 12.6 shows the cost comparison for different concretes used in this research.

The results show that for M60 grade concrete, the difference in the experimental results of modulus of elasticity is 8.82%, 15.99%, and 13.92% for conventional concrete, as per IS 456:2000, ACI-209, and BS 8110:1997, respectively. The difference of modulus of elasticity noted by IS 456:2000, ACI-209, and BS 8110:1997 for concrete containing waste is 3.97%, 13.38%, and 17.93%, respectively.

The results show that for M40 grade concrete, the difference in the experimental results of modulus of elasticity is 9.23%, 14.39%, and 10.52% for conventional concrete, as per IS 456:2000, ACI-209, and BS 8110:1997, respectively. The difference of modulus of elasticity noted by IS 456:2000, ACI-209, and

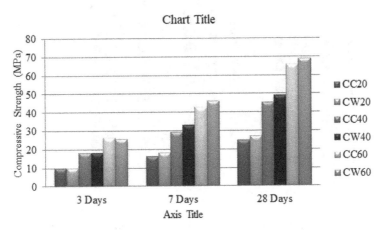

FIGURE 12.3 Compressive strength variation of different concrete mixes at different ages.

**TABLE 12.5**

**Comparison of Experimental Modulus of Elasticity (MPa) with the Values Obtained from Different Standards**

| | | Modulus of Elasticity (Es) (MPa) | | | |
|---|---|---|---|---|---|
| Concrete | $f_c$ (MPa) | Experimental | IS 456:2000 | ACI–209 | BS 8110:1997 |
| CC20 | 24.68 | 24358.57 | 24839.48 | 25998.41 | 23029.60 |
| C20W50 | 26.86 | 25432.05 | 25913.32 | 27722.33 | 23509.20 |
| CC40 | 45.28 | 30801.73 | 33645.21 | 35236.54 | 27561.60 |
| C40W50 | 49.15 | 33539.22 | 35053.53 | 37003.93 | 28413.00 |
| CC60 | 65.68 | 37236.46 | 40521.60 | 43193.58 | 32049.60 |
| C60W50 | 68.94 | 39929.68 | 41515.06 | 45273.52 | 32766.80 |

BS 8110:1997 for concrete containing waste is 4.51%, 10.33%, and 15.28%, respectively.

The results show that for M20 grade concrete, the difference in the experimental results of modulus of elasticity is 1.97%, 6.73%, and 5.45% for conventional concrete, as per IS 456:2000, ACI-209, and BS 8110:1997, respectively. The difference of modulus of elasticity as noted by IS 456:2000, ACI-209, and BS 8110:1997 for concrete containing waste is 1.89%, 9.0%, and 7.56%, respectively.

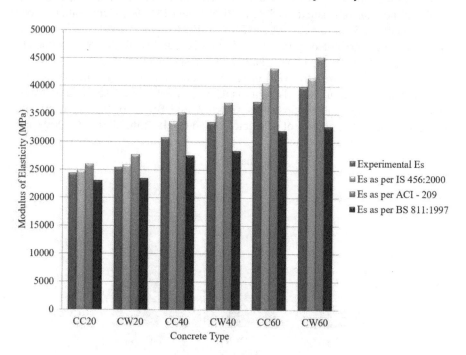

**FIGURE 12.4** Comparison of experimental investigated modulus of elasticity for various concrete mixes with the values obtained as per different standards.

**TABLE 12.6**
**Cost Comparison**

| Concrete | Material | Cement | Waste Powder | Sand | 10 mm Aggregate | 20 mm Aggregate | Admixture | Total Amount in INR | Savings in INR per m³ of Concrete |
|---|---|---|---|---|---|---|---|---|---|
| CC20 | Quantity (kg) | 300 | - | 696 | 400 | 1029 | - | | 527.25 |
| | Price (INR) | 1800 | - | 661.2 | 260 | 771.75 | - | 3492.95 | |
| CW20 | Quantity (kg) | 150 | 150 | 906 | 445 | 669 | 2.7 | | |
| | Price (INR) | 900 | 225 | 860.7 | 289.25 | 501.75 | 189 | 2965.7 | |
| CC40 | Quantity (kg) | 465 | - | 531 | 423 | 728 | - | | 656.7 |
| | Price (INR) | 2790 | - | 504.45 | 274.95 | 546 | - | 4115.4 | |
| CW40 | Quantity (kg) | 235 | 235 | 680 | 428 | 740 | 3.1 | | |
| | Price (INR) | 1410 | 352.5 | 646 | 278.2 | 555 | 217 | 3458.7 | |
| CC60 | Quantity (kg) | 510 | - | 687 | 441 | 664 | 3.18 | | 1018.8 |
| | Price (INR) | 3060 | - | 652.65 | 286.65 | 498 | 222.6 | 4719.9 | |
| CW60 | Quantity (kg) | 265 | 265 | 648 | 464 | 696 | 3.92 | | |
| | Price (INR) | 1590 | 397.5 | 615.6 | 301.6 | 522 | 274.4 | 3701.1 | |

Water charges are not considered in this cost comparison.

## 12.4 CONCLUSION

Through experiments, it was determined that steel industry waste powder, grade 120 as per SAI, provides the desired compressive strength for 50% cement replacement in concrete. It is recommended to use this concrete made with industrial waste to increase the strength of concrete structures, but not in places that require high early strength. Using waste improves workability and decreases the water/binder ratio due to the increase in paste volume caused by the lower density of waste. The economic benefits are also important for using industrial waste as cement replacement. The cost comparison shows the cost of CW20, CW40, and CW60 is lower by INR. 527.25 per $m^3$, INR. 656.70 per $m^3$, and INR. 1018.80 per $m^3$ compared to CC20, CC40, and CC60, respectively.

The concrete containing waste has higher strength, is more economical, and reduces the use of cement, directly affecting $CO_2$ production. It can be concluded that this concrete containing waste offers a sustainable solution for the construction industry.

## REFERENCES

ACI Committee 209. 1999. "Prediction of creep and shrinkage, and temperature effects in concrete structures." *American Concrete Institute.*

ASTM Standards. 2002. "Standard test method for static modulus of elasticity and Poisson's ratio of concrete in compression." ASTM C 469-02. United States.

ASTM Standards. 2004. "Standard specification for slag cement for use in concrete and mortars." ASTM C 989-04. United States.

British Standards. 1992. "Specification for ground granulated blast furnace slag for use with Portland cement." BS 6699:1992.

British Standards Institution. 1985. "Structural use of concrete — Part 2: Code of practice for special circumstances." 8110-2:1985. London.

Bureau of Indian Standard. 2002. "Specification for coarse and fine aggregate from natural sources for concrete." IS 383:1970.

Bureau of Indian Standard. 2005. "Plain and reinforced concrete code of practice." IS 456:2000. New Delhi, India.

Bureau of Indian Standard. 2013. "Ordinary Portland Cement, 53 Grade - Specifications." IS 12269:2013.

Dimitrova, S. 1995. "Metal sorption on blast-furnace slag." *Water Research* 30(1): 228–232. https://doi.org/10.1016/0043-1354(95)00104-S.

Fais, B., N. Sabio, and N. Strachan. 2016. "The critical role of the industrial sector in reaching long-term emission reduction, energy efficiency and renewable targets." *Applied Energy* 162: 699–712. doi: 10.1016/j.apenergy.2015.10.112.

Govindan, K., K. Madan Shankar, and K. Devika. 2016. "Sustainable material selection for construction industry – A hybrid multi criteria decision making approach." *Renewable and Sustainable Energy Reviews* 55: 1274–1288. doi: 10.1016/j.rser.2015.07.100.

Hannant, D. J. 1967. "Strain behavior of concrete up to 95°C under compressive stresses." Proceeding Conference on Pre-stressed Concrete Pressure Vessels, Group C, Institution of Civil Engineers. 57–71. London.

Kan, T., V. Strezov, T. Evans, and P. Nelson. 2015. "Trace element deportment and particle formation behaviour during thermal processing of iron ore: technical reference for risk assessment of iron ore processing." *Journal of Cleaner Production* 102: 384–393. doi: 10.1016/j.jclepro.2015.04.032.

Motz, H., and J. Geiseler. 2001. "Products of steel slags an opportunity to save natural resources." *Waste Management* 21: 285–293. doi: 10.1016/S0956-053X(00)00102-1.

Nasser, K. W., and A. M. Neville. 1965. "Creep of concrete at elevated temperatures." *Journal of American Concrete Institute* 62(12): 1567–1579.

Ozturk, Z., and E. Gultekin. 2015. "Preparation of ceramic wall tiling derived from blast furnace slag." *Ceramics International* 41: 12020–12026. doi: 10.1016/j.ceramint.2015.06.014.

Proske, T., S. Hainer, M. Rezvani, and C. Graubner. 2013. "Eco-friendly concretes with reduced water and cement contents — Mix design principles and laboratory tests." *Cement and Concrete Research* 51: 38–46. doi: 10.1016/978-0-12-804524-4.00004-X.

Rajamane, N. P., J. Annie Peter, J. K. Dattatreya, M. Neelamegam, and S. Gopalakrishnan. 2003. "Improvement in properties of high performance concrete with partial replacement of cement by ground granulated blast furnace slag." *The Institution of Engineers (India)* IE (I): 38–42.

Sadek, D. 2014. "Effect of cooling technique of blast furnace slag on the thermal behavior of solid cement bricks." *Journal of Cleaner Production* 79: 134–141. doi: 10.1016/j.jclepro.2014.05.033.

Swamy, R., and Bouikni, A. 1990. "Some engineering properties of slag concrete as influenced by mix proportioning and curing." *Aci Materials Journal* 87: 210–220.

Swamy, R. N. 2008. "Sustainable concrete for the 21st century concept of strength through durability." *Japan Society of Civil Engineers Concrete Committee Newsletter* 13: 1–28.

Zainuddin, A. M, and Norlinda R. Wan. 2017. "Advances in Process Integration research for $CO_2$ emission reduction - a review." *Journal of Cleaner Production* 167: 1–13. doi: 10.1016.j.jclepro.2017.08.138.

Zhao, H., W. Sun, X. Wu, and B. Gao. 2015. "The properties of the self-compacting concrete with fly ash and ground granulated blast furnace slag mineral admixtures." *Journal of Cleaner Production* 95: 66–74. doi: 10.1016.j.jclepro.2015.02.050.

# 13 A Study of Environmental Disclosure Practices in India with an Emphasis on the Mining Sector

*Sumona Ghosh*

## CONTENTS

## 13.1 INTRODUCTION

Disclosures of environmental issues are gaining momentum in the corporate world. Both natural disasters like landslides, pollution, and others as well as human-made environmental disasters such as the Union Carbide's methyl isocyanate gas leak in Bhopal are challenging the world. These have created greater awareness among various groups of stakeholders such as the public at large, employees, contractors, suppliers, customers, and officials who have different objectives and expectations. Increasing media exposure on the diminishing global supplies of natural resources and on environmental disasters have forced societies to direct their attention toward businesses accepting more responsibility for their actions (Gray, Owen and Maunders 1988). Tilt (1997) suggested that through environmental reporting and disclosures, stakeholders can get the necessary assurance that corporations would be held accountable for their environmental activities. Effective environmental reporting thus plays a crucial part today.

Environmental reporting is the practice of measuring, disclosing, and communicating to internal and external stakeholders about the impact of the organization's actions on the environment as well as organizational environmental performance (FEE, 2000; GRI, 2006) so that sustainable development goals can be achieved (GRI, 2006). Environmental reporting information is important and required by the public for decision-making with respect to investment, lending, consumption, and labor supply (Deegan and Rankin 1999). Companies that fail to pay heed to environmental issues could incur losses with respect to customers' confidence, and legal costs and heavy damages for the environmental problems that they have caused (Blacconiere and Patten 1994).

Environmental information can be disseminated through an annual report (Gray, Kouhy, and Lavers 1995), the company website (Tagesson et al. 2009), a sustainability report (GRI, 2006) or as a stand-alone report. The environmental report can be a voluntary report (Gray, Owen, and Adams 1996), required by the stakeholders (Van Der Laan 2004), or a mandatory report provided to governmental agencies (Fallan and Fallan 2009). Environmental reporting can be done either by following a structured approach (based on environmental reporting frameworks such as Global Reporting Initiative (GRI) or it can be an unstructured approach, such as voluntary self-reporting.

The focus of this study will be on analyzing and comparing the level of annual report disclosure with respect to environmental activities for selected companies in India for the time period from 2014–2015 to 2018–2019 using a constructed disclosure index. The reason for focusing on this time period and area is that with the enforcement of Section 135 of the Companies Act, 2013 (MCA, 2013), India became the first country to include social responsibility issues (environment being an important component) in company law and make expenditure regarding such issues mandatory for corporations based on pre-specified criteria. This is still a voluntary exercise for the rest of the world and it is left to the discretion of the corporation. Compulsory reporting of such socially responsive issues including environmental activities undertaken by corporations has become mandatory in India.

Schedule VII to the Companies Act of 2013 specifies activities that companies may include in their environment policies and one such activity that the act has emphasized among others is "ensuring environmental sustainability – ecological balance – protection of flora and fauna – Animal welfare – Agro forestry – Conservation of natural resources – Maintaining quality of soil, air and water including contribution to the Clean Ganga Fund set up by the Central Government for rejuvenation of river Ganga." Besides Business Responsibility Report (BRR), a report mandated by SEBI (2015) for top 500 listed entities and prepared according to the specified format contains nine principles. Out of the nine principles to be disclosed in the BRR, the sixth principle talks about the disclosure an organization needs to make to show its environmental awareness, measures taken to protect the environment, and the efforts made to restore the environment.

## 13.2 LITERATURE REVIEW

Studies on environmental reporting practices were not very prominent before the 1970s although research was carried out in the area of environmental accounting practices in different corporations. We will therefore approach this segment from three perspectives – an account of the general studies on environmental reporting and disclosure practices globally from 1980s to 2018, environmental reporting and disclosure index studies, and studies on environmental reporting and the mining sector.

### 13.2.1 AN ACCOUNT OF THE GENERAL STUDIES ON ENVIRONMENTAL REPORTING AND DISCLOSURE PRACTICES GLOBALLY FROM THE 1980s TO 2018

Annual reports of companies have been examined by various researches to analyze the association of a company's environmental disclosure as well as environmental performance, the quality and accuracy of environmental reporting, factors affecting environmental disclosure (Ingram and Frazier 1980, Wiseman 1982, Shane and Spicer 1983, Guthrie and Parker 1989, Wu, Liu, and Sulkowski 2010, Mooney et al. 2011, Joshi and Suwaidan 2011, Galani, Gravas and Stavropoulos 2011, Roy and Ghosh 2011). Rao, Tilt, and Lester (2012) through their research observed a strong positive correlation between the extent of environmental reporting and the proportions of independent and female directors on a board. There were many studies which looked into the disclosure practices of the companies and their studies revealed that the level of environmental disclosure was low (Savage 1994, Deegan and Gordon 1996, Deegan and Rankin 1996, Lemon and Cahan 1997, Neu, Warsame and Pedwell 1998, Choi 1999). During the 2000s, most of the social and environmental disclosure studies that were conducted indicated poor quality social and environmental disclosure (Belal 2000, Belal 2001, Belal 2008, Imam 2000, Moneva and Liena 2000, Tilt 2001, Gray and Bebbington 2001, Hughes et al. 2001, Fortes 2002, Sahay 2004, Elijido-Ten 2004, Cowan and Gadenne 2005, Sarumpaet 2005, Hossain, Islam, and Andrew 2006, Yusoff, Lehman and Nasir 2006, Kamla 2007, Rizk et al. 2008, Malarvizhi and Yadav 2008, Brammer and Pavelin 2008, Singh and Joshi 2009, Pahuja 2009, Ousama and Fatima 2010, Clarkson et al. 2011, Sen, Mukherjee and Pattanayak 2011, Liu et al. 2011, Cuesta and Valor 2013, Said et al. 2013, Bowrin 2013, He and Loftus 2014, Ahmad and Hossain 2015, Nurhayati et al. 2016).

### 13.2.2 ENVIRONMENTAL REPORTING AND DISCLOSURE INDEX STUDIES

Many studies have constructed a disclosure index in order to examine the quality and quantity of environmental reporting in the annual reports. One of the earliest indices found in the literature is that of "Wiseman's environmental disclosure index (EDI)," which contains 18 items (Fekrat, Inclan and Petroni 1996, Cormier and Gordon 2001, Buhr and Freedman 2001). This index is believed to better represent the quality of environmental information disclosed in annual reports (Buhr and Freedman 2001, Wiseman 1982, Freedman and Wasley 1990, Walden and Schwartz 1997, Stagliano and Walden 1998, Bewley and Li 2000, Hughes, Anderson, and Golden 2001, Patten 2002, Clarkson et al. 2008, Galani, Gravas and Stravropoulous 2011, Asaolu et al. 2011, Eljayash 2015, Fekrat, Inclan, and Petroni 1996, Cormier and Magnan 1999, Cormier and Gordon 2001, Cormier and Magnan 2003, Cormier, Magnan, and Van 2005, Bakhtiar 2005, Ten 2009).

### 13.2.3 STUDIES ON ENVIRONMENTAL REPORTING AND THE MINING SECTOR

Core sector which includes mining operations is considered as one of the most polluting sectors since it involves a high degree of environmental impacts. According to several authors (Roe and Samuel 2007, Schueler, Kuemmerle, and Schröder 2011, Amponsah-Tawiah and Dartey-Baah 2011), mining activities have tremendously affected the key elements of the environment (land, water, and air) and have endangered the health of local people. Some studies have indicated that the quantity and quality of environmental disclosure with respect to mining companies is extremely low (Hutomo 1995, Christopher, Cullen, and Soutar 1998, Yongvanich and Guthrie 2004, Günther, Hoppe, and Poser 2007, Perez and Sanchez 2009, Sen, Mukherjee, and Pattanayak 2011, Ane 2012).

A review of the literature shows that most environmental disclosure studies have focused their attention on developed countries, with less attention paid to developing countries. In terms of sector, we observe that there are very limited studies examining environmental disclosure in the mining sector, especially with respect to India with its enormous mining sector. This study attempts to overcome the observed limitations of the literature by investigating the environmental disclosure practices of the mining companies in developing countries with special reference to India.

## 13.3   OBJECTIVES

Mining activities in emerging economies have tremendously affected land, water, and air, resulting in serious consequences for the health of local people. There are various negative effects of the mining industry on the environment, with water pollution, soil erosion, formation of sinkholes, and loss of biodiversity being among the worst effects. The adverse impact that mining operations have on the environment is the reason why it is categorized under the red category (industrial sectors having a pollution index of 60 and above) according to Indian Ministry of Forest and Environment.

In spite of the fact that the mining sector is a very sensitive sector we observe that there have been very limited studies on this topic, especially with respect to India. Environmental disclosure practices are more robust in some parts of the developed world, but for the developing world it remains a new methodology. Given this background, the present study focuses on the analysis of the level of annual report reporting with respect to the environmental activities for selected companies in India for the time period from 2014–2015 to 2018–2019 using a constructed disclosure index. It is also concerned with the following objectives:

- *Objective 1:* To construct a Corporate Environmental Reporting Index (CERI) for different time periods across all companies and to differentiate between the companies on that basis.

- *Objective 2:* To investigate the extent to which the companies in India are disclosing their environmental activities and responsibilities in their reporting by using the constructed CERI.
- *Objective 3:* To ascertain the factors emerging out of the parameters defined by GRI4 in the context of environmental reporting for different periods.
- *Objective 4:* To estimate the functional relationship between the factors and the CERI for different time periods in order to examine the extent of influence of the factors on the CERI.

## 13.4  METHODOLOGY

### 13.4.1  DATA SOURCE AND STUDY DESIGN

An empirical and analytical study of environmental reporting was made for the financial years 2014–15 to 2018–19. The study used secondary sources whereby each selected company's annual reports, websites, and sustainability reports were analyzed. Data were generated from the analyses using Longitudinal Qualitative Document Analysis for the five-year period. The GRI environmental index was used to identify and classify the environmental disclosures. The most commonly used but prominent global framework for voluntary corporate environmental reporting is the GRI Sustainability Reporting Guidelines. Since the GRI Index also includes elements of other frameworks such as ISO 14000, and the Global·Sullivan Principles, it is the most comprehensive framework. In this analysis, the G4 Guidelines (Appendix A) were employed to analyze the level of environmental reporting from 2014–2015 to 2018–2019. We identified and calculated the GRI environmental performance elements that the companies disclosed in their annual reports/sustainability reports to measure the level of environmental reporting in each year. If the GRI element was present, we assigned a value of 1 and in its absence we assigned a value of 0.

### 13.4.2  EXTENT OF RESPONSIVENESS TO ENVIRONMENT DISCLOSURE

The extent of responsiveness to environment disclosure for the financial years 2014–15 to 2017–18 was analyzed, and four main stages of responsiveness to environment disclosure have been identified in this chapter – high, medium, progressive, and low. We considered those companies to be highly responsive/responsiveness level is low to environment disclosures whose environment reporting score was higher/lower than the grand average score for each of the five years. We considered companies to be progressive in their disclosures if their score gradually improved, and the remaining companies were regarded to be in the moderate stage.

### 13.4.3  SELECTION OF COMPANIES

The Bombay Stock Exchange of India (BSE) and the National Stock Exchange of India (NSE) are the two major stock exchanges in India. The BSE was Asia's first stock exchange, established in 1875. It is the world's 11th largest stock exchange.

The number of companies publicly listed on the BSE exceeds 5,500 companies. The NSE, the world's 12th largest stock exchange, was founded in 1992 and began trading in 1994 as the first demutualized electronic exchange in India. Our study is based on firms listed on the BSE and NSE indices as of August 6, 2019. We have studied all the firms belonging to the sector "mining and minerals" as classified by the BSE. This gave us a set of 30 firms. The annual reports and sustainability reports of these companies were analyzed for the time period from 2014–15 to 2018–19, i.e., a five-year time period.

### 13.4.4 METHOD

Longitudinal Qualitative Document Analysis was employed for the period from 2014–15 to 2018–19. The CERI was constructed by examining the annual reports and sustainability reports of any of the 30 companies belonging to the "mining and minerals" sector. The CERI was calculated in the following manner:

$$CERI = \left(\sum di/n\right) \times 100 = (TS/n) \times 100 \qquad (13.1)$$

where:

CERI = Corporate Environmental Reporting Index
di: 1 if item i is disclosed; 0 if item i is not disclosed
n = no. of items = maximum score
TS = total score

40 items of disclosure as provided by the G4 Guidelines have been used in this study. For each sampled company the maximum score would be 40. "100" is the maximum CERI score and "0" is the minimum score. Companies whose disclosure level is high would have a score of 100 or closer to it and companies whose disclosure level is low would have a score of "0" or closer to it. We then identified the four levels of responsiveness - high, medium, progressive, and low.

The variables for factor analysis were obtained through Longitudinal Qualitative Document Analysis. Exploratory Factor Analysis (EFA) was applied for extraction of factors. Reliability test was carried out using Cronbach's alpha. Multiple regression analysis was performed in order to determine the impact of the independent variables, i.e., factors so generated from factor analysis on CERI. This estimation of multiple regression models has been done for all the five years under study.

## 13.5   RESULTS AND DISCUSSIONS

Table 13.1A focused on the individual CERI of the companies for the time period 2014–2019. The study revealed an average CERI of 16.13% which was extremely low. Table 13.1B showed the extent to which the companies were responsive to environmental disclosure and differentiated into high level, low level, progressive level and moderate level of responsiveness to environmental disclosure.

## TABLE 13.1A
## Individual CERI of the Companies for the Time Period 2014–2015 to 2018–2019

| Name | 2014–15 | 2015–16 | 2016–17 | 2017–18 | 2018–19 | EORE |
|---|---|---|---|---|---|---|
| Coal India | 40% | 60% | 62.50% | 67.50% | 60% | H |
| Vedanta | 65% | 65% | 67.50% | 67.50% | 67.50% | H |
| Moil | 15% | 20% | 20% | 15% | 15% | M |
| Guj Mineral | 45% | 47.50% | 47.50% | 47.50% | 47.50% | H |
| Maithan Alloys | 7.50% | 7.50% | 7.50% | 7.50% | 7.50% | L |
| Sandur Manganes | 7.50% | 12.50% | 10% | 10% | 10% | L |
| India Metals | 22.50% | 22.50% | 17.50% | 15% | 20% | P |
| Orissa Minerals | 7.50% | 7.50% | 7.50% | 5.00% | 7.50% | L |
| Deccan Gold | 5% | 5% | 5% | 5% | 5% | L |
| Ashapura Mines | 7.50% | 12.50% | 10% | 8% | 10% | L |
| 20 Microns | 20% | 17.50% | 17.50% | 17.50% | 20.00% | H |
| RSMML | 17.50% | 17.50% | 17.50% | 17.50% | 17.50% | H |
| Indsil Hydro | 2.50% | 2.50% | 2.50% | 2.50% | 2.50% | L |
| Raw Edge Industrial Solutions Ltd | 7.50% | 7.50% | 7.50% | 7.50% | 7.50% | L |
| Ferro Alloys | 5% | 7.50% | 5% | 5% | 5% | L |
| Shyam Centuries | 7.50% | 7.50% | 7.50% | 7.50% | 7.50% | L |
| Shirpur Gold | 10% | 10% | 10% | 10% | 10% | L |
| Nagpur Power | 2.50% | 2.50% | 2.50% | 0.00% | 0.00% | L |
| Facor Alloys | 5% | 2.50% | 2.50% | 2.50% | 7.50% | L |
| Impex Ferro Tech | 5% | 5% | 5% | 5% | 5% | L |
| NMDC | 40% | 42.50% | 42.50% | 42.50% | 45% | H |
| Hindustan Zinc Ltd | 40% | 70% | 70% | 70% | 70% | H |
| Ceeta Industries Ltd | 2.50% | 2.50% | 2.50% | 2.50% | 2.50% | L |
| Glittek Granite Ltd | 2.50% | 2.50% | 2.50% | 2.50% | 2.50% | L |
| Kiocl Ltd | 27.50% | 25% | 25% | 32.50% | 32.50% | H |
| Foundry Fuel Products Ltd | 2.50% | 2.50% | 2.50% | 2.50% | 2.50% | L |
| ASI Industries Ltd | 7.50% | 7.50% | 7.50% | 7.50% | 7.50% | L |

*(Continued)*

**TABLE 13.1A (*Continued*)**
**Individual CERI of the Companies for the Time Period 2014–2015 to 2018–2019**

| Name | 2014–15 | 2015–16 | 2016–17 | 2017–18 | 2018–19 | EORE |
|---|---|---|---|---|---|---|
| Pacific Industries Ltd | 2.50% | 5% | 5% | 5% | 5% | L |
| Oriental Trimex Ltd | 2.50% | 2.50% | 2.50% | 2.50% | 2.50% | L |
| Inani Marbles and Industries Ltd | 0 | 0 | 0 | 0 | 0 | L |
| **Average CERI** | **14.42%** | **16.67%** | **16.42%** | **16.33%** | **16.83%** | |
| **Average CERI (5 years )** | **16.13%** | | | | | |

*Note:* (a) EORE refers to extent of responsiveness to environmental disclosure; (b) H: refers to high level of responsiveness to environmental disclosure; (c) M: refers to moderate level of responsiveness to environmental disclosure; (d) L: refers to low level of responsiveness to environmental disclosure; (e) P: refers to progressive level of responsiveness to environmental disclosure.

From Table 13.1B we observe that environmental disclosure level was very high among 8 companies, environmental disclosure level was moderate in 1 company, environmental disclosure level was progressive in 1 company and environmental disclosure level was very low among 20 companies.

Factor analysis was carried out for the years 2014–2015 to 2018–2019. The summarized results have been presented in Table 13.2A. Principal Components Analysis (PCA) is employed here. From our analysis, all variables with values below 0.5 have been excluded. All factors with an eigenvalue greater than unity have been retained. This resulted in nine factors that keep 86.99% of the total variance in the original data for the year 2014–2015, eight factors that keep 86.33% of the total variance in the original data for the year 2015–2016, nine factors that keep 88.58% of the total variance in the original data for the year 2016–2017, nine factors that keep 89.44% of the total variance in the original data for the year 2017–2018, nine factors that keep 87.69% of the total variance in the original data for the year 2018–2019, and these factors characterize the dimensionality of our 40 individual indicators. 'Varimax' rotation was used to rotate the reduced solution. This method allowed the retained factors to be correlated to strengthen the interpretability of the PCA solution.

There are 40 items/variables with respect to environmental disclosures (refer to appendix). These have been grouped into six areas namely, pollution and waste management, sustainable use of resources, general environmental policy, climate change, environmental management, and protection of biodiversity. This has been presented

**TABLE 13.1B**

**Classification of Companies by Extent of Responsiveness to Environment Disclosure for the Time Period 2014–2019**

| High Level of Responsiveness to Environment Disclosure | Moderate Level of Responsiveness to Environment Disclosure | Progressive Responsiveness to Environment Disclosure | Low Level of Responsiveness to Environment Disclosure |
|---|---|---|---|
| Coal India | MOIL | India metals | Maithan Alloys |
| Vedanta | | | Sandur Manganes |
| Guj Mineral | | | Orissa Minerals |
| | | | |
| 20 Microns | | | Deccan Gold |
| RSMML | | | Ashapura Mines |
| NMDH | | | Indsil Hydro |
| Hindustan Zinc Ltd | | | Raw Edge Industrial Solutions Ltd |
| KIOCL Ltd | | | Ferro Alloys |
| | | | Shyam Centuries |
| | | | Shirpur Gold |
| | | | Nagpur Power |
| | | | Facor Alloys |
| | | | Impex Ferro Tech |
| | | | Ceeta Industries Ltd |
| | | | Glittek Granite Ltd |
| | | | Foundry Fuel Products Ltd |
| | | | ASI Industries Ltd |
| | | | Pacific Industries Ltd |
| | | | Oriental Trimex Ltd |
| | | | Inani Marbles and Industries Ltd |

in Table 13.2B. In Table 13.2C, we have summarized the variables associated with each factor for all the years.

Cronbach's $\alpha$, a measure of scale reliability, is computed since scale reliability of those dimensions that make up the factors is essential. Cronbach's $\alpha$ is a measure of internal consistency that ranges from 0 to 1. The mean (median) of the alpha coefficient is 0.99 and it is consistent with the benchmarks suggested by Nunnally (1967).

Our aim was to understand the effect of various environmental disclosure variables on the CERI. The independent variables are various environmental disclosure aspects provided by the GRI framework. Hence we examine whether there exists an association or not assuming that a linear relationship exists between them. Using the factor scores generated by PCA we run regression. Models I and 2 both yielded R-square values of 99.6%, respectively. This signified that 99.6% of the variability in CERI is accounted for by the models considering the predictor variables. Similarly,

**TABLE 13.2A**

**Factors, Factor Loadings, and Variances**

| ENVTDI | 2014–15 | 2015–16 | 2016–17 | 2017–18 | 2018–19 |
|---|---|---|---|---|---|
| Cumulative Variance (%) | 86.99 | 86.33 | 88.58 | 89.44 | 87.69 |
| **Factor Loadings** | | | | | |
| EP1 | FACTOR2(.548) | FACTOR1(.690) | FACTOR1(.951) | FACTOR1(.962) | FACTOR1(.958) |
| EP2 | FACTOR2(.698) | FACTOR2(.917) | FACTOR1(.811) | FACTOR1(.763) | FACTOR1(.787) |
| EP3 | FACTOR9(.573) | FACTOR6(.642) | FACTOR5(.779) | FACTOR3(.575) | FACTOR3(.456) |
| EP4 | FACTOR3(.996) | FACTOR1(.915) | FACTOR1(.776) | FACTOR1(.863) | FACTOR1(.834) |
| EP5 | FACTOR3(.996) | FACTOR5(.992) | FACTOR3(.991) | FACTOR6(.984) | FACTOR8(.982) |
| EP6 | FACTOR4(.877) | FACTOR4(.890) | FACTOR2(.842) | FACTOR3(.841) | FACTOR2(.807) |
| EP7 | FACTOR3(.996) | FACTOR5(.992) | FACTOR3(.991) | FACTOR2(.633) | FACTOR9(.807) |
| EP8 | FACTOR5(.777) | FACTOR3(.770) | FACTOR9(.728) | FACTOR8(.731) | FACTOR4(.761) |
| PWM1 | FACTOR5(.579) | FACTOR3(.743) | FACTOR1(.739) | FACTOR1(.738) | FACTOR1(.724) |
| PWM2 | FACTOR5(.671) | FACTOR3(.676) | FACTOR1(.568) | FACTOR1(.619) | FACTOR1(.608) |
| PWM3 | FACTOR5(.511) | FACTOR3(.632) | FACTOR6(.733) | FACTOR2(.805) | FACTOR3(.684) |
| PWM4 | FACTOR5(.442) | FACTOR7(.465) | FACTOR7(.587) | FACTOR2(.514) | FACTOR6(.457) |
| PWM5 | FACTOR4(.785) | FACTOR6(.602) | FACTOR2(.503) | FACTOR3(.552) | FACTOR3(.854) |
| PWM6 | FACTOR6(.743) | FACTOR6(.638) | FACTOR5(.774) | FACTOR1(.741) | FACTOR1(.659) |
| SUR1 | FACTOR8(.557) | FACTOR8(.442) | FACTOR5(.444) | FACTOR2(.461) | FACTOR3(.579) |
| SUR2 | FACTOR7(.690) | FACTOR1(.893) | FACTOR1(.951) | FACTOR1(.962) | FACTOR1(.958) |
| SUR3 | FACTOR2(.575) | FACTOR2(.709) | FACTOR4(.599) | FACTOR1(.710) | FACTOR2(.540) |
| SUR4 | FACTOR9(.787) | FACTOR1(.383) | FACTOR4(.427) | FACTOR9(.635) | FACTOR9(.616) |
| SUR5 | FACTOR8(.811) | FACTOR8(.761) | FACTOR7(.816) | FACTOR5(.769) | FACTOR6(.852) |

|  |  |  |  |  |  |
|---|---|---|---|---|---|
| SUR6 | FACTOR4(.648) | FACTOR4(.664) | FACTOR2(.747) | FACTOR3(.753) | FACTOR2(.647) |
| SUR7 | FACTOR7(.798) | FACTOR1(.786) | FACTOR8(.889) | FACTOR7(.912) | FACTOR7(.895) |
| CC1 | FACTOR1(.867) | FACTOR1(.758) | FACTOR1(.886) | FACTOR1(.869) | FACTOR1(.879) |
| CC2 | FACTOR1(.867) | FACTOR1(.893) | FACTOR1(.951) | FACTOR1(.962) | FACTOR1(.958) |
| CC3 | FACTOR1(.579) | FACTOR1(.626) | FACTOR1(.834) | FACTOR2(.621) | FACTOR4(.555) |
| CC4 | FACTOR1(.704) | FACTOR1(.758) | FACTOR1(.862) | FACTOR1(.869) | FACTOR1(.879) |
| CC5 | FACTOR2(.924) | FACTOR1(.893) | FACTOR1(.951) | FACTOR1(.962) | FACTOR1(.958) |
| **BIODIVERSITY** | FACTOR5(.579) | FACTOR3(.736) | FACTOR1(.713) | FACTOR2(.633) | FACTOR1(.563) |
| EM1 | FACTOR4(.743) | FACTOR1(.645) | FACTOR1(.703) | FACTOR1(.685) | FACTOR1(.672) |
| EM2 | FACTOR4(.722) | FACTOR4(.818) | FACTOR2(.922) | FACTOR3(.910) | FACTOR2(891) |
| EM3 | FACTOR2(.793) | FACTOR2(.522) | FACTOR1(.707) | FACTOR1(.618) | FACTOR1(.640) |
| EM4 | FACTOR6(.493) | FACTOR7(.529) | FACTOR2(.592) | FACTOR2(.663) | FACTOR4(.499) |
| EM5 | FACTOR3(.996) | FACTOR5(.992) | FACTOR3(.991) | FACTOR6(.984) | FACTOR8(.982) |
| EM6 | FACTOR6(.725) | FACTOR3(.334) | FACTOR6(.870) | FACTOR2(.839) | FACTOR3(.519) |
| EM7 | FACTOR1(.867) | FACTOR2(.632) | FACTOR1(.811) | FACTOR1(.962) | FACTOR1(.787) |
| EM8 | FACTOR1(.522) | FACTOR1(.572) | FACTOR1(.788) | FACTOR1(.778) | FACTOR1(.787) |
| EM9 | FACTOR1(.910) | FACTOR1(.915) | FACTOR2(.621) | FACTOR1(.649) | FACTOR2(.739) |
| EM10 | FACTOR6(.569) | FACTOR2(.823) | FACTOR4(.02) | FACTOR1(.613) | FACTOR1(.586) |
| EM11 | FACTOR2(.924) | FACTOR2(.851) | FACTOR1(.902) | FACTOR4(.886) | FACTOR5(.872) |
| EM12 | FACTOR3(.996) | FACTOR7(.666) | FACTOR1(.616) | FACTOR4(.886) | FACTOR5(.872) |
| EM13 | FACTOR8(.584) | FACTOR8(.672) | FACTOR7(.749) | FACTOR5(.731) | FACTOR6(.675) |

*Note:* (a) ENVTDI refers to items related to environmental disclosures. (b) The results of the rotated component matrix are shown here. Extraction method: principal component analysis. The rotation method used is Varimax with Kaiser Normalization. Rotation converged in 10 iterations in 2014–15, 12 in 15–16, 10 iterations in 2016–17, 15 iterations in 17–18, and 35 iterations in 2018–19.

**TABLE 13.2B**

**Groupings of the Various Items of Environmental for the Time Period 2014–2019**

| General Environmental Policy | Pollution and Waste Management | Sustainable Use of Resources | Climate Change | Protection of Biodiversity | Environmental Management |
|---|---|---|---|---|---|
| EP1 | PWM1 | SUR1 | CC1 | Biodiversity | EM1 |
| EP2 | PWM2 | SUR2 | CC2 | | EM2 |
| EP3 | PWM3 | SUR3 | CC3 | | EM3 |
| EP4 | PWM4 | SUR4 | CC4 | | EM4 |
| EP5 | PWM5 | SUR5 | CC5 | | EM5 |
| EP6 | PWM6 | SUR6 | | | EM6 |
| EP7 | | SUR7 | | | EM7 |
| EP8 | | | | | EM8 |
| | | | | | EM9 |
| | | | | | EM10 |
| | | | | | EM11 |
| | | | | | EM12 |
| | | | | | EM13 |

*Source:* Authors own research.

**TABLE 13.2C**

**Variables Associated with Each Factor for the Time Period 2014–2019**

| Year | Factors | Variables Included | Name of the Factor |
|---|---|---|---|
| 2014–15 | 1 | CC1,CC2,CC3,CC4,EM7,EM8,EM9 | Climate change and environmental management |
| | 2 | EP1,EP2,SUR3,CC5,EM3,EM11 | General environmental policy, sustainable use of resources, climate change, environmental management, |
| | 3 | EP4,EP5,EP7,EM5,EM12 | General environmental policy and environmental management |
| | 4 | EP6,PWM5,SUR6,EM1,EM2 | General environmental policy, sustainable use of resources, environmental management |
| | 5 | EP8,PWM1,PWM2,PWM3,PWM4, BIODIVERSITY | General environmental policy, pollution and waste management, biodiversity |
| | 6 | PW6,EM4,EM6,EM10 | Pollution and waste management, environmental management |
| | 7 | SUR2,SUR7 | Sustainable use of resources |

**TABLE 13.2C (*Continued*)**
**Variables Associated with Each Factor for the Time Period 2014–2019**

| Year | Factors | Variables Included | Name of the Factor |
|---|---|---|---|
| | 8 | SUR1,SUR5,EM13 | Sustainable use of resources, environmental management |
| | 9 | EP3,SUR4 | General environmental policy, sustainable use of resources |
| 2015–16 | 1 | EP1,EP4,SUR4,SUR7,CC1,CC2,CC3, CC4,CC5,SUR2,EM1,EM8,EM9 | General environmental policy, sustainable use of resources, climate change, environmental management |
| | 2 | EP2,SUR3,EM3,EM7,EM10,EM11 | General environmental policy and environmental management |
| | 3 | EP8,PWM1,PWM2,PWM3, BIODIVERSITY,EM6 | General environmental policy, pollution and waste management, biodiversity, environmental management |
| | 4 | EP6,SUR6,EM2 | General environmental policy, sustainable use of resources, environmental management |
| | 5 | EP5,EP7,EM5 | General environmental policy and environmental management |
| | 6 | EP3,PWM5,PWM6,EM12 | General environmental policy, pollution and waste management, environmental management |
| | 7 | PWM4,EM4 | Pollution and waste management, environmental management |
| | 8 | SUR1,SUR5,EM13 | Sustainable use of resources, environmental management |
| 2016–17 | 1 | EP1,EP2,EP4,PWM1, PWM2,SUR2, CC1,CC2,CC3,CC4,CC5, BIODIVERSITY,EM1,EM3, EM4,EM7,EM8 | General environmental policy, sustainable use of resources, climate change, environmental management, pollution and waste management, biodiversity |
| | 2 | EP6,PWM5,SUR6,EM2,EM9 | General environmental policy, sustainable use of resources, environmental management, pollution and waste management |
| | 3 | EP5,EP7,EM5 | General environmental policy and environmental management |
| | 4 | SUR3,SUR4,EM10,EM11,EM12 | Sustainable use of resources, environmental management |
| | 5 | EP3,PWM6,SUR1 | General environmental policy, sustainable use of resources, pollution and waste management |
| | 6 | PWM3,EM6 | Pollution and waste management, environmental management |

*(Continued)*

**TABLE 13.2C (*Continued*)**
**Variables Associated with Each Factor for the Time Period 2014–2019**

| Year | Factors | Variables Included | Name of the Factor |
|---|---|---|---|
| | 7 | PWM4,SUR5,EM13 | Sustainable use of resources, pollution and waste management, environmental management |
| | 8 | SUR7 | Sustainable use of resources |
| | 9 | EP8 | General environmental policy |
| 2017–18 | 1 | EP1,EP2,EP4,PWM1,PWM2,PWM6, SUR2,SUR3,CC1,CC2,CC4,CC5, EM1,EM3,EM7,EM8,EM9,EM10 | General environmental policy, sustainable use of resources, climate change, environmental management, pollution and waste management |
| | 2 | EP3,EP7,PWM3,PWM4,SUR1,CC3, BIODIVERSITY,EM4,EM6 | General environmental policy, sustainable use of resources, climate change, environmental management, pollution and waste management, biodiversity |
| | 3 | EP6,PWM5,SUR6,EM2 | General environmental policy, sustainable use of resources, environmental management, pollution and waste management |
| | 4 | EM11,EM12 | Environmental management |
| | 5 | SUR5,EM13 | Sustainable use of resources, environmental management |
| | 6 | EP5,EM5 | General environmental policy and environmental management |
| | 7 | SUR7 | Sustainable use of resources |
| | 8 | EP8 | General environmental policy |
| | 9 | SUR4 | Sustainable use of resources |
| 2018–19 | 1 | SUR2,CC5,CC2,EP1,CC1,CC4,EP4, EM7,EP2,EM8,PWM1,PWM6,EM1, EM3,PWM2,EM10,BIODIVERSITY | General environmental policy, sustainable use of resources, climate change, environmental management, pollution and waste management, biodiversity |
| | 2 | EM2,EP6,EM9,SUR6,SUR3 | General environmental policy, sustainable use of resources, environmental management |
| | 3 | PWM5,PWM3,SUR1,EM6,EP3 | General environmental policy, sustainable use of resources, environmental management, pollution and waste management |
| | 4 | EP8,CC3,EM4 | General environmental policy, climate change, environmental management |
| | 5 | EM12,EM11 | Environmental management |

## TABLE 13.2C (*Continued*)
## Variables Associated with Each Factor for the Time Period 2014–2019

| Year | Factors | Variables Included | Name of the Factor |
|---|---|---|---|
| | 6 | SUR5,EM13,PWM4 | Sustainable use of resources, pollution and waste management, environmental management |
| | 7 | SUR7 | Sustainable use of resources |
| | 8 | EM5,EP5 | General environmental policy and environmental management |
| | 9 | EP7,SUR4 | General environmental policy, sustainable use of resources |

Models 3, 4, and 5 yielded R-square values of 99.7%, 99% and 99.8% respectively. In all the models all the factors turned out to be significant. We have presented the results in Table 13.3.

## TABLE 13.3
## Results of the Regression Analysis

| Model | 1(2014–15) | 2(2015–16) | 3(2016–17) | 4(2017–18) | 5(2018–19) |
|---|---|---|---|---|---|
| | CERI | CERI | CERI | CERI | CERI |
| Climate change and environmental management | .000 | | | | |
| General environmental policy, sustainable use of resources, climate change, environmental management, | .000 | .000 | | | |
| General environmental policy and environmental management | .000 | .000 | .000 | .000 | .000 |
| General environmental policy, sustainable use of resources, environmental management | .000 | .000 | | | .000 |
| General environmental policy, pollution and waste management, biodiversity | .000 | | | | |
| Pollution and waste management, environmental management | .000 | .000 | .000 | | |
| Sustainable use of resources | .000 | | .000 | .000 | .000 |
| Sustainable use of resources, environmental management | .000 | .000 | .000 | .000 | |
| General environmental policy, sustainable use of resources | .000 | | | | .000 |

(*Continued*)

**TABLE 13.3 (*Continued*)**
**Results of the Regression Analysis**

| Model | 1(2014–15) CERI | 2(2015–16) CERI | 3(2016–17) CERI | 4(2017–18) CERI | 5(2018–19) CERI |
|---|---|---|---|---|---|
| General environmental policy, pollution and waste management, biodiversity, environmental management | | .000 | | | |
| General environmental policy, pollution and waste management, environmental management | | .000 | | | |
| General environmental policy, sustainable use of resources, climate change, environmental management, pollution and waste management, biodiversity | | | .000 | .000 | .000 |
| General environmental policy, sustainable use of resources, pollution and waste management | | | .000 | | |
| Sustainable use of resources, pollution and waste management, environmental management | | | .000 | | .000 |
| General environmental policy, pollution and waste management | | | | | |
| General environmental policy, sustainable use of resources, climate change, environmental management, pollution and waste management | | | | .000 | |
| General environmental policy, sustainable use of resources, environmental management, pollution and waste management | | | .000 | .000 | .000 |
| Environmental management | | | | .000 | .000 |
| General environmental policy | | | .000 | .000 | |
| General environmental policy, climate change, environmental management | | | | | .000 |
| $R^2$ | .996 | .996 | .997 | .999 | .998 |
| CONSTANT | 14.417(0.000) | 16.638 (0.000) | 16.417 (0.000) | 16.350 (0.000) | 16.750 (0.000) |

## 13.6    CONCLUSION

An exploratory study was undertaken to understand the relationship between various environmental disclosure variables and CERI. We assembled a detailed data set comprising 30 listed companies belonging to the mining sector. We ensured that data for specified five years and with all firm year data was available on the selected variables. Our focus was to study and analyze the level of disclosure of environmental activities in the annual reports of the companies belonging to the mining sector of India for the time period from 2014–2015 to 2018–2019 using a constructed disclosure index. The reason for choosing the mining sector is due to the fact that mining activities have tremendously affected the environment (land, water, and air) leading to serious consequences for the health of locals.

The study revealed an average CERI of 16.13% which happened to be extremely low. 26.67% of the companies chosen for our analysis, proved to be highly responsive to environmental disclosures, 3.33% of the companies chosen for our analysis showed moderate level of responsiveness to environmental disclosures, 3.33% of the companies chosen for our analysis proved to be progressive in nature with respect to environmental disclosures, and 66.67% of the companies chosen for our analysis, proved to be having an extremely low level of responsiveness to environmental disclosures. Hence, environmental reporting needs its promotion among such companies in the Indian context so that these companies can also stand on the same footing as those who have shown very high responsiveness to environmental disclosures.

Our analysis have shown that *General environmental policy, environmental management,* and *Sustainable use of resources* for all the years have turned out to be significant, having a significant and positive impact on CERI.

Environmental policy is defined as "any action deliberately taken to manage human activities with a view to prevent, reduce, or mitigate harmful effects on nature and natural resources, and ensuring that man-made changes to the environment do not have harmful effects on humans or the environment" (McCormick 2001). Public perceptions of a company's commitment to environmental protection and sustainable development can be positively influenced by corporations publishing their environmental policy statements. This could result in an increased market share and improved stakeholder relations. Companies belonging to the mining sector as per our study have shown very poor responsiveness to the development and reporting of environment policy. Hence, effective disclosure which has such a significant impact on CERI needs to be encouraged.

Environmental management is the management of human interactions with, and impact on, the environment. Environmental management is concerned with monitoring environmental changes, with attempts to increase benefits to humans, and to reduce the degradation of the environment from human activities.

The sustainable use of natural resources is using the natural resources in a way and at a rate that benefits the entire human community and does not degrade the resilience of ecosystems. Resources are the backbone of every economy. Capital stocks are built up by using and transforming resources, which in turn builds up the wealth of present and future generations. However the manner in which these scarce resources are being used is detrimental to the welfare of developing countries and may induce serious damage that goes beyond the carrying capacity of the environment.

The majority of the companies in our study have failed to disclose the manner in which they are dealing with sustainable use of natural resources and what steps they are undertaking to prevent these scarce resources from being depleted, in spite of the fact that sustainable use of natural resources has such a significant impact on CERI.

Our study focused on environmental disclosure practices in India with a special emphasis on the mining sector; it is hoped that future research will advance empirical work in this important area of study.

## APPENDIX A

### Environmental Disclosure Index

| | |
|---|---|
| **ENVTDV** | |
| EP1 | Company efforts to take into account environmental issues |
| EP2 | Appropriate, assessments or environmental certifications |
| EP3 | Employee training programs on environmental protection. |
| EP4 | Budget dedicated to environmental protection and environmental risk mitigation |
| EP5 | Financial provisions for environmental risk and pollution. |
| EP6 | R&D expenditures for pollution abatement |
| EP7 | Financing for pollution control equipment or facilities |
| EP8 | The implementation of HSE (Health Safety Environment) approach |
| PWM1 | Measures to prevent, reduce, or compensate for air |
| PWM2 | Measures to prevent, reduce, or compensate for water |
| PWM3 | Measures to prevent, reduce, or compensate for soil emissions severely affecting the environment. |
| PWM4 | Measures to prevent, recycle, and dispose of waste |
| PWM5 | Taking into account noise and other forms of pollution specific to activity |
| PWM6 | Significant environmental impacts of transporting products and other goods and materials used for the organization's operations, and transporting members of the workforce |
| SUR1 | Water use and water supply based on local constraints |
| SUR2 | Percentage and total volume of water recycled and reused |
| SUR3 | The consumption of raw materials and steps taken to improve their efficient use |
| SUR4 | Energy consumption |
| SUR5 | Measures to improve energy efficiency |
| SUR6 | Percentage of renewable energy used. |
| SUR7 | Land use. |
| CC1 | Greenhouse gas emissions |
| CC2 | Adaptation to climate change impacts |
| CC3 | Initiatives to reduce greenhouse gas emissions and reductions achieved. |
| CC4 | NOx, SOx, and other significant air emissions by type and weight |
| CC5 | Total number and volume of significant spills |
| **BIODIVERSITY** | Protection of biodiversity |
| EM1 | Presence of Independent verification/assurance about environmental information disclosed |
| EM2 | Joint projects with other firms on environmental management |
| EM3 | Goals and targets |

| EM4 | Certification ISO 14000 |
| EM5 | Participation in elaborating of environmental standards |
| EM6 | The existence of environmental department or office for pollution control |
| EM7 | Existence of terms and conditions applicable to suppliers and/or customers regarding environmental practices |
| EM8 | A statement about the firm's compliance with existed environmental regulation or other schemes (GRI, UNGC, ISO26000) |
| EM9 | Environmental litigation/lawsuits against the company |
| EM10 | External environmental awards, prizes/or inclusion in sustainability index |
| EM11 | Commitments to an environmental or sustainable development charter |
| EM12 | Extra financial environmental rating by Sustainability Rating Agency |
| EM13 | Participation in environmental association/initiatives to improve environmental practices |

*Source:*    Based on G4 guidelines.

# REFERENCES

Ahmad, N. N. and Hossain, D. M. 2015. Climate change and global warming discourses and disclosures in the corporate annual reports: A study on the Malaysian companies. *Procedia – Social and Behavioral Sciences*, 172(27), 246–253.

Amponsah-Tawiah, K. and Dartey-Baah, K. 2011. The mining industry in Ghana: A blessing or a curse. *International Journal of Business and Social Science*, 2(12), 62–70.

Ane, P. 2012. An assessment of the quality of environmental information disclosure of corporation in China. *Systems Engineering Procedia*, 5, 420–426.

Asaolu, T. O., Agboola, A. A., Ayoola, T. J. and Salawu, M. K. 2011. Sustainability reporting in the Nigerian, oil and gas sector. Proceedings of the Environmental Management Conference, Federal University of Agriculture, Abeokuta, Nigeria. pp. 1–24.

Bakhtiar, A. 2005. Comparing environmental reporting practice of public listed companies in Malaysia in 1999 and 2003: An investigation of quantity and quality. Master dissertation, International Islamic University Malaysia.

Belal, A. 2000. Environmental reporting in developing countries: Empirical evidence from Bangladesh. *Eco-Management and Auditing*, 7(3), 114–121

Belal, A. 2001. A study of corporate social disclosures in Bangladesh. *Managerial Auditing Journal*, 16(5), 274–289.

Belal, A. R. 2008. *Corporate Social Responsibility Reporting in Developing Countries: The Case of Bangladesh*. Aldershot: Ashgate.

Bewley, K. and Li, Y. 2000. Disclosure of environmental information by Canadian manufacturing companies: A voluntary disclosure perspective. *Advances in Environmental Accounting and Management*, 1, 201–226.

Blacconiere, W.G. and Patten, D.M. 1994. Environmental disclosures, regulatory costs and changes in firm value. *Journal of Accounting and Economics*, 18(3), 357–77.

Bowrin, A. R. 2013. Corporate social and environmental reporting in the Caribbean. *Social Responsibility Journal*, 9(2), 259–280.

Brammer, S. and Pavelin, S. 2008. Factors influencing the quality of corporate environmental disclosure. *Business Strategy and the Environment*, 17(2), 120–136.

Buhr, N and Freedman, M. 2001. Institutional factors and differences in environmental disclosure between Canada and the United States. *Critical Perspective Account*, 1(2/3), 293–322.

Choi, Jong-Seo. 1999. An Investigation of the initial voluntary environmental disclosures made in Korean semi-annual financial reports. *Pacific Accounting Review*, 11(1), June. Available at SSRN: https://ssrn.com/abstract=177210

Christopher, T., Cullen, L., and Soutar, G. 1998. Australian mining companies environmental disclosure. *Accountability and Performance*, 4(10), 17–41.

Clarkson, P. M., Li, Y., Richardson, G. D. and Vasvari, F. P. 2008. Revisiting the relation between environmental performance and environmental disclosure: an empirical analysis. *Accounting, Organizations and Society*, 33(4/5), 303–327.

Clarkson, Peter. M., Overell, Michael. B. and Chapple, Larelle. 2011. Environmental reporting and its relation to corporate environmental performance. *Abacus*, 4(1), 27–60.

Cormier, D, Gordon, I. M. 2001. An examination of social and environmental reporting strategies. *Account Audit Account*, 14(5), 587–616.

Cormier, D. and Gordon, J. M. 2001. An examination of social and environmental reporting strategies. *Accounting, Auditing & Accountability Journal*, 14(5), 587–616.

Cormier, Denis and Magnan, Michel, 2003. Environmental reporting management: a continental European perspective. *Journal of Accounting and Public Policy*, Elsevier, 22(1), 43–62.

Cormier, D., Magnan, M. 1999. Corporate environmental disclosure strategies: Determinants, costs and benefits. *Journal of Accounting, Auditing and Finance*, 14(3), 429–451.

Cormier, D, Magnan, M. and Van, Velthoven Barbara. 2005. Environmental disclosure quality in large German companies: Economic incentives, public pressures or institutional conditions? *European Accounting Review* 14(1):3–39.

Cowan, S. and Gadenne, D. 2005. Australian corporate environmental reporting: A comparative analysis of disclosure practices across voluntary and mandatory disclosure system. *Journal of Accounting & Organizational Change*, 1(2), 165–79.

Cuesta, M. and Valor, C. 2013. Evaluation of the environmental, social and governance information disclosed by Spanish listed companies. *Social Responsibility Journal*, 9(2), 220–240.

Deegan, Craig and Gordon, Ben. 1996. A study of the environmental disclosure practices of Australian corporations. *Accounting and Business Research*, 26(3), 187–199.

Deegan, C. and Rankin, M. 1996. Do Australian companies report environmental news objectively? An analysis of environmental disclosures by firms prosecuted successfully by the Environmental Protection Authority. *Accounting, Auditing & Accountability Journal*, 9(2), 50–67.

Deegan, C. and Rankin, M. 1999. The environmental reporting expectation gap: Australian evidence. *British Accounting Review*, 31, 313–46.

Dintimala, Y. and Amril, T. A. 2018. The effect of ownership structure, financial and environmental performances on environmental disclosure. *Accounting Analysis Journal*, 7(1), 70–77.

Elijido-Ten, E. 2004. Determinants of environmental disclosure in a developing country: An application of stakeholder theory, paper presented to the Asia Pacific Interdisciplinary Research in Accounting (APIRA), Singapore.

Eljayash, Kamal Mehemed. 2015. Documentation of environmental disclosure practices in the oil companies in the countries of the Arab Spring – some evidences from Egypt, Libya and Tunisia. *Journal of Economics, Business and Management*, 3(10), 954–960.

Ezeagba, C. E., Rachael, J.-A. C. and Chiamaka, U. 2017. Environmental accounting disclosures and financial performance: A study of selected food and beverage companies in Nigeria (2006–2015). *International Journal of Academic Research in Business and Social Sciences*, 7(9), 162–174.

Fallan, E. and Fallan, L. 2009. Voluntarism versus regulation: Lessons from public disclosure of environmental performance information in Norwegian companies. *Journal of Accounting & Organisation Change*, 5(4), 472–489.

FEE, Fédération des Experts Comptables Européens, 2000. Towards a generally accepted framework for environmental reporting, July

Fekrat, M. A., Inclan, C. and Petroni, D. 1996. Corporate environmental disclosure: Competitive disclosure hypothesis using 1991 annual report data. *The International Journal of Accounting*, 31(2), 175–95

Fortes, H. (2002). The need for environmental reporting by companies. *Greener Management International*, Winter, 77–92.

Freedman, M. and Wasley, C. 1990. The association between environmental performance and environmental disclosure in annual reports and 10Ks. *Advances in Public Interest Accounting*, 3, 183–193.

Galani, D., Gravas, E. and Stavropoulos, A. 2011. Company characteristics and environmental policy. *Business Strategy and the Environment*, 21, 236–247.

Grace, O. N. and Ndubuisi, O. 2018. International financial reporting standards (IFRS) disclosure and performance of NIGERIA listed companies. *Cogent Business & Management*, 5:1.

Gray R. H. and Bebbington, K.J. 2001. *Accounting for the Environment* 2nd Edition, London: Sage.

Gray, R., Kouhy, R. and Lavers, S. 1995. Corporate social environmental reporting: a review of the literature and a longitudinal study of UK disclosure. *Accounting, Auditing & Accountability Journal*, 8(2), 78–101.

Gray, R., Owen, D. and Adams, C. 1996. *Accounting and Accountability; Changes and Challenges in Corporate Social and Environmental Reporting*, Harlow: Prentice Hall Europe.

Gray, R. H., Owen, D. L and Maunders, K. T. 1988. Corporate social reporting: emerging trends in accountability and the social contract. *Accounting, Auditing & Accountability Journal*, 1(1), 6–20.

GRI - Global Reporting Initiative. 2006. Sustainability reporting guidelines., Amsterdam, available online at www.globalreporting.org

Günther, E., Hoppe, H. and Poser, C. 2007. Environmental corporate social responsibility of firms in the mining and oil and gas industries: Current status quo of reporting following GRI guidelines. *Greener Management International*, 53, 7–25.

Guthrie, J.E. and Parker, L.D. 1989. Corporate social reporting: a rebuttal of legitimacy theory. *Accounting and Business Research*, 9(76), 343–352.

He, C. and Loftus, J. 2014. Does environmental reporting reflect environmental performance? Evidence from China. *Pacific Accounting Review*, 26 (1/2), 134–154.

Hossain, M., Islam, K. and Andrew, J. 2006. Corporate social and environmental disclosure in developing countries: evidence from Bangladesh. *Proceedings of the Asian Pacific Conference on International Accounting Issues*, Hawaii. University of Wollongong, Research Online, Faculty of Commerce - Papers Faculty of Commerce

Hughes, S. B., Anderson, A., and Golden, S. 2001. Corporate environmental disclosures: Are they useful in determining environmental performance? *Journal of Accounting and Public Policy*, 20, 217–240.

Hutomo, Y. B. S. 1995. Voluntary environmental disclosure by Australian listed mineral mining companies: An application of stakeholder theory (Unpublished master dissertation). Edith Cowan University, Joondalup, Western Australia.

Imam, S. 2000. Corporate social performance reporting in Bangladesh. *Managerial Auditing Journal*, 15(3): 133–141.

Ingram, R. W. and Frazier, K. B. 1980. Environmental performance and corporate disclosure. *Journal of Accounting Research* (Autumn), 614–621.

Innocent, O. C. and Gloria, O. T. 2018. Firm attributes and corporate environmental performance: Evidence from quoted industrial firms on Nigerian stock exchange. *Scholars Journal of Economics, Business and Management*, 5(9), 854–863

Joshi, P. L. and Gao, S. 2009. Multinational corporations; corporate social and environ-
    mental disclosures (CSED) on web sites. *International Journal of Commerce and
    Management*, 19 (1), 27–44.
Joshi, P. L. and Suwaidan, M. S. 2011. Determinants of environmental disclosures by Indian
    industrial listed companies: Empirical study. *International Journal of Accounting and
    Finance*, 3(2), 109–130.
Kamla, R. 2007. Critically appreciating social accounting and reporting in the Arab Middle
    East: a postcolonial perspective. *Advances in International Accounting*, 20, 105–177.
Lemon, A. J. and Cahan, S. F. 1997. Environmental Legislation and Environmental Disclosures:
    Some Evidence from New Zealand. *Asian Review of Accounting*, 5(1), 78–105.
Liu, Z., Liu, T., McConkey, B. and Li, X. 2011. Empirical analysis on environmental disclo-
    sure and environmental performance level of listed steel companies. *Energy Procedia*,
    5, 2211–2218.
Malarvizhi, P. and Yadav, S. 2008. Corporate Environmental Reporting on the Internet an
    Insight into Indian practices, Paper presented at the 11th Annual Convention of the
    Strategic Management Forum, May 8–10, Indian Institute of Technology, Kanpur,
    pp. 1–14.
MCA, 2013, Companies Act, 2013, Ministry of Corporate Affairs, Government of India, New
    Delhi. http://www.mca.gov.in/Ministry/pdf/CompaniesAct2013.pdf
McCormick, M. J. 2001. Self-efficacy and leadership effectiveness: applying social cognitive
    theory to leadership. *Journal of Leadership Studies*, 8(1):22–33.
Moneva, Jose M. and Liena, Fernando. 2000. Environmental disclosures in the annual reports
    of large companies in Spain, *European Accounting Review*, 9(1), 7–29.
Mooney, S. D. et al. 2011. Late quaternary fire regimes of Australasia. *Quaternary Science
    Reviews*, 30, 28–46
Neu, D., Warsame, H., and Pedwell, K. 1998. Managing public impressions: Environmental
    disclosures in annual reports. *Accounting, Organizations and Society*, 23(3), 265–282.
Nunnally, J. 1967. *Psychometric Theory*. New York, NY: McGraw Hill.
Nurhayati, R., Taylor, G., Rusmin, R., Tower, G. and Chatterjee, B. 2016. Factors determin-
    ing social and environmental reporting by Indian textile and apparel firms: A test of
    legitimacy theory. *Social Responsibility Journal*, 12(1), 167–189.
Ousama, A. A. and Fatima, A. H. 2010. Voluntary disclosure by Shariah approved companies:
    An exploratory study. *Journal of Financial Reporting and Accounting*, 8(1), 35–49.
Pahuja, Shuchi. 2009. "Relationship between environmental disclosures and corporate char-
    acteristics: a study of large manufacturing companies in India. *Social Responsibility
    Journal*, 5(2), 227–244.
Patten, D. M. 2002. The relation between environmental performance and environmental
    disclosure: A research note. *Accounting, Organizations and Society*, 27(8), 763–773.
Perez, F. and Sanchez, L. E. 2009. Assessing the evolution of sustainability reporting in the
    mining sector. *Environmental Management*, 43, 949–61.
Rao, K. K., Tilt, C. A. and Lester, L. H. 2012. "Corporate governance and environmental
    reporting: an Australian study. *Corporate Governance*, 12(2), 143–163.
Rizk, R., Dixon, R., Woodhead, A. 2008. Corporate social and environmental reporting: A
    survey of disclosure practices in Egypt. *Social Responsibility Journal*, 4(3), 306–323.
Roe, A. and Samuel, J. 2007. *The Challenge of Mineral Wealth: Using Resource Endowments
    to Foster Sustainable Development: Ghana, Case Country*. International Council of
    Mining and Metals, London.
Roy, A. andGhosh, S. 2011. The bilateral association between discretionary environmental
    disclosure quality and economic performance: An Asian perspective. *The IUP Journal
    of Accounting Research & Audit Practices*, 10(2), 7–27.
Sahay, A. 2004, Environmental reporting by Indian corporations. *Corporate Social
    Responsibility and Environmental Management*, 11(1), 12–22.

Said, R., Omar, N. and Abdullah, W. N. 2013. Empirical investigations on boards, business characteristics, human capital and environmental reporting. *Social Responsibility Journal*, 9(4), 534–553.

Sarumpaet, S. 2005. The relationship between environmental performance and financial performance of Indonesian companies. *Jurnal Akuntasi & Keuangan*, 7 (2), 89–98

Savage, A. A. 1994. Corporate social disclosure practices in South Africa: research note. *Social and Environmental Accounting*, 14(1).

Schueler, V., Kuemmerle, T., Schröder, H. 2011. Impacts of surface gold mining on land use systems in Western Ghana. *Ambio*, 40(5), 528–539.

SEBI (Listing Obligations and Disclosure Requirements) Regulations, 2015.

Sen, M., Mukherjee, K. and Pattanayak, J. K. 2011. Corporate environmental disclosure practices in India. *Journal of Applied Accounting Research*, 12(2), 139–156.

Shane, P. and Spicer, B. 1983. Market response to environmental information produced outside the firm. *The Accounting Review*, 521–38

Singh, G. and Joshi, M. 2009. Environment management and disclosure practices of Indian companies. *International Journal of Business Research*, 9(2), 116–128.

Stagliano, A. J. and Walden, W. D. 1998. Assessing the quality of environmental disclosure themes. Conference Paper No. 75, APIRA 98, Osaka, Japan

Tagesson, T., Blank, V., Broberg, P. and Collin, S. O. 2009. What explains the extent and content of social and environmental reporting in Swedish listed corporations. *Corporate Social Responsibility and Environmental Management*, 16, 352–364.

Ten, E. E. 2009. Can stakeholder theory add to our understanding of Malaysian environmental reporting attitudes?" *Malaysian Accounting Review* 8(2), 85–110

Tilt C. A. 2001. The content and disclosure of Australian corporate environmental policies. *Accounting, Auditing and Accountability Journal*, 14(2), 190–212.

Tilt, C. A. 1997.Environmental policies of major companies: Australian evidence. *British Accounting Review*, 29, 367–394.

Van Der Laan, S. 2004. The role of theory in explaining motivation for corporate social disclosures: Voluntary disclosures vs 'Solicited' disclosures. Paper presented at the Fourth Asia Pacific Interdisciplinary Research in Accounting Conference, Singapore.

Walden, W. and Schwartz, B. 1997. Environmental disclosures and public policy pressure'. *Journal of Accounting and Public Policy*, 16, 125–154.

Wiseman, J. 1982. An evaluation of environmental disclosure made in annual reports. *Accounting, Organizations and Society*, 7(1), 53–63.

Wu, Jia, Liu, Linxiao, Sulkowski, Adam. 2010. Environmental disclosure, firm performance, and firm characteristics: An analysis of S&P 100 firms. *Journal of Academy of Business and Economics* 10(4):73–83.

Yongvanich, K. and Guthrie, J. 2004. Extended performance reporting: An examination of the Australian mining industry. *Accounting Forum*, 29(1), 103–119.

Yusoff, H., Lehman, G. and Nasir, N. M. 2006. Environmental engagements through the lens disclosure practices: a Malaysian story. *Asian Review of Accounting*, 14(2), 122–148.

# Section V

## Green Healthcare

# 14 Identifying Green Physicians

## A Segmentation Study in Eastern India

*Supti Mandal*

## CONTENTS

## 14.1  BACKGROUND

Environmental concerns have increased drastically due to the many major ecological catastrophes that have occurred over recent decades. The concern is that
although Earth's population is increasing exponentially, natural resource supplies
are finite. Thus, the environment in economically developing countries, such as
India and China, have been particularly threatened by problems like poverty, pollution, overpopulation, and environmental depletion and degradation due to rapid
industrialization.

We have to think of something new, as the old ways are no longer adequate. We
have to put emphasis on our values using a paradigm that prioritizes the partnership between humankind and Earth. To even survive, humans must have harmony
with the Earth, its resources, and its biosphere. This is just the beginning, and
there is much work to do as time ticks away. Few Indian companies have started
the journey. Marketers, due to their pivotal position, need to lead the drive by
ensuring a profit for their companies that is not at the cost of destroying the Earth
and its biosphere.

### 14.1.1  GREEN MARKETING

Environmental marketing emerged in the 1970s when the concept of ecological marketing (Kinnear et al. 1974) arose from the demands of ecologically concerned consumers (Kardash 1974).

People perceive green marketing as the promotion or advertising of products with
environmental characteristics. However, in general, green marketing is much more
holistic in essence and can be applied to consumer goods, industrial goods, and
even services. Few researchers consider environmental marketing as "a component
of *sustainable marketing*, a concept that aims to create customer value, social value
and ecological value" (Belz 2006).

Polonsky (1994) proposed green marketing as "all activities designed to generate
and facilitate any exchanges intended to satisfy human needs or wants, such that the
satisfaction of these needs and wants occurs, with minimal detrimental impact on
the natural environment." Green marketing can be termed as a shift in managerial
thought from viewing the natural and physical environment as an external influence
on decision making to viewing it as central to marketing and management strategy
(Menon and Menon 1997).

#### 14.1.1.1  Green Marketing in India

Although some Indians and some urban consumers are to some extent aware of the
benefits of going green, this is still an emerging concept for Indian masses. There
needs to be a lot of time and effort invested to educate consumers about environmental issues.

In a research study, Saxena and Khandelwal (2008) found that industries in India
are quite willing to adapt to the green marketing philosophies for future sustainable

development. According to Shrikanth and Raju (2012), Indian companies are venturing into green marketing for multiple reasons:

- India has a diverse and fairly sizable green market
- Many Indian companies believe that achieving environmental objectives may help reach profit objectives, so these companies have started to behave in an environmentally friendly fashion
- The Indian government's newly framed legislation has made adopting green marketing a compulsion
- Many proactive companies have taken up green marketing to maintain their competitive edge

The most efficient companies who perform better in combating climate changes are in Europe, the United States of America, and Japan. There are also five Indian companies (Essar Oil, Larsen & Tubro, Tech Mahindra, Tata Consultancy Service, and Wipro) on the global list (CDP Climate Performance Leadership Index 2014). The list has 187 companies across the globe, proving that a low-carbon future does not mean low profit (Mohani 2014).

The report "Indian Companies Decouple Business Growth from Carbon Disclosure Project" (2014) analyzed the top 200 Indian companies by market capitalization, and concluded that Indian companies are quite willing to collaborate with the government to keep themselves in line with regulatory changes (Menon 2014).

## 14.1.2 Pharmaceutical Industry

As defined by the US Census Bureau, the pharmaceutical industry comprises companies engaged in researching, developing, manufacturing, and distributing drugs for human or veterinary use. The pharmaceutical industry is among the largest and most dynamic industrial sectors. The global market for pharmaceuticals is anticipated to grow from around $1 trillion in 2015 to $1.3 trillion by 2020, representing an annual growth rate of 4.9%. Indian pharmaceutical sector is characterized by a high level of competitiveness, more limited economic budgets, expensive sales, and promotion activities. In this context, it is becoming increasingly important to understand which factors influence doctors in prescribing medicines in order to design appropriate marketing strategies.

The Indian pharmaceutical sector is highly organized, and this market is expected to expand at a compound annual growth rate (CAGR) of 23.9% to reach US$55 billion by 2020. As far as the technology, quality, and vast range of medicines are concerned, India holds an important position in the pharmaceutical realm, especially among other economically developing countries.

### 14.1.2.1 Environmental Effect of Pharmaceuticals

Although there is an extensive global campaign to protect the environment, there is hardly any talk about pharmaceutical pollution. Pharmaceutical industries release very toxic contaminants directly into the environment or after chemically modifying them. Activated sludge, or similar kinds of wastewater treatment methods, are not sufficiently effective in completely removing the active pharmaceutical ingredients and other wastewater constituents from wastewater. Pharmaceutical effluents, especially near the pharmaceutical industrial zones, directly or indirectly affect the environment, human health, and animal health, as well as drinking water.

### 14.1.3 MARKET SEGMENTATION

In this complex market, it is important to understand how doctors can be categorized, in order to design appropriate marketing strategies. In the pharmaceutical sector, the primary customer is the general practitioner because they prescribe medicines, so enterprise profits cannot be obtained without considering customer satisfaction. Different bases have been chosen by previous researchers to appropriately segment pharmaceutical markets.

#### 14.1.3.1 Demographic and Environmental Criteria

The demographic data that was included for doctors were age, gender, religion, marital status, education, workplace, income, length of practice, specialization of physicians, and the number of patients per day that the doctor treats.

*14.1.3.1.1 Environmental Attitude*

Hogg and Vaughan (2005, 150) defined "attitude" as "a relatively enduring organization of beliefs, feelings, and behavioral tendencies toward socially significant objects, groups, events or symbols." A number of behavioral researchers have documented attitudes to predict behavioral intention and explanatory factors of variants in individual behavior. (Kotchen and Reiling 2000).

*14.1.3.1.2 Environmental Concerns*

Considering attitude as a variable, Kinnear et al. (1974) mentioned that green consumers must express environmental concern. The New Environmental Paradigm (NEP) scale was originally designed by Dunlap and Van Liere in 1978. Later, in 2000, Dunlap et al. revised the NEP scale to accommodate its inherent multidimensionality, offer a balanced set of pro- and anti-NEP items, and avoid outmoded terminology. In environmental psychology literature, NEP is "commonly taken to measure general environmental concern" (Poortinga et al. 2004).

*14.1.3.1.3 Perceived Consumer Effectiveness*

As first described by Kinnear et al. (1974), perceived consumer effectiveness (PCE) is a measure of the extent to which an individual believes that he or she can be effective in pollution abatement. PCE as "the evaluation of the self in the context of the issue" (Berger and Corbin 1992, 80–81) is different from "an attitude that reflects an evaluation of an issue" (Tesser and Shaffer 1990).

*14.1.3.1.4 Environmental Knowledge*

In consumer research, knowledge is recognized as a characteristic that influences all phases in the decision process. Environmental knowledge is defined by Fryxell and Carlos (2003) as "a general knowledge of facts, concepts and relationships concerning the natural environment and its major ecosystems." In the research of Larasati and Hananto (2012), environmental knowledge is found to be a significant predictor of consumers' green purchase behavior.

### 14.1.3.2  Value Orientation

Recent psychological theories and studies on values are based on the work of Stern (1999) and, more recently, Schwartz (1992, 1994). Schwartz (1992) defines "value" as "a desirable trans situational goal varying in importance, which serves as a guiding principle in the life of a person or other social entity."

Egoistic value orientation includes two value types: power and achievement. Social altruistic value orientation comprises two value types: universalism and benevolence. The motivational goal for biospheric value orientation is the welfare of the biosphere (Attfield 1981; Naess 2003).

Environmental values are values held by people with regard to the relationship between humans and the natural environment, typically viewed as a continuum ranging from purely anthropocentric to purely biocentric or purely ecocentric (Dunlap and Van Liere Kent, 1978; Edwards-Jones et al. 2000).

## 14.2  LITERATURE REVIEW

In the United States, marketers quickly determined the scope of green market and profiled it.. The landmark study was undertaken by The Roper Organization on behalf of S.C. Johnson. In Roper Organisation I (1990), the entire US population was divided into five segments: True Blue Greens (11%), Greenback Green (11%), Sprouts (26%), Grousers (24%), and Basic Brown (28%). Roper Organisation II (1993) points out that in spite of the 1990 economic recession, the environmental concern still held the interest of American consumers. In the same year, Cambridge Reports (1993) published their report, "Green Consumerism," and produced a fairly basic four part segmentation based on a clustering of behavior and attitudinal responses: dedicateds (34%), avid environmental consumers (34%), temperates (16%), and do nothings (16%). Said (1997) identified six segments of the Australian market according to its greenness: Living Greens (28.5%), Nurturing Greens (21%), Grudging Greens (12%), Lip Service Greens (17%), Light Brown Battlers (12.5%), and Brown Bombers (9%). Laroche and Bergeron (2001) investigated the demographic, psychological, and behavioral profiles of Canadian consumers who are willing to pay more for environmentally friendly products. This segment of consumers is more likely to be female, married, and with at least one child living at home. Jain and Kaur (2004) conducted their study, "Green Marketing: An Attitudinal and Behavioural Analysis of Indian Consumers," in Delhi. The results reveal that respondents demonstrate a rise in environmental concerns and perceive people in India as willing to take environmentally friendly steps to reduce environmental problems. Zografos and Allcroft (2007) carried out a market segmentation study based on the environmental values of potential ecotourists, and their results indicate that demand for ecotourism is not confined to ecocentric segments. Mostafa (2007) showed that Egyptian men are more environmentally concerned and have a more positive outlook toward green purchases compared with women, contradicting the previous study conducted in the West. Saxena and Khandelwal (2008) found Indian consumers to have a strong positive attitude toward green marketing. Cegarra-Navarro et al. (2009) demonstrated the relevance of environmental knowledge in creating customer capital and explored how organizations can create environmental knowledge through "green communities" in the Spanish pharmaceutical industry.

Finisterra do Paço et al. (2009) showed that consumers who buy green products do so for specific reasons, and environmental and demographic variables have significance in green buying.

Ham (2009) profiled an average green consumer in the Republic of Croatia. The empirical study of Tantawi et al. (2009) contradicts the traditional beliefs that environmental concern is a luxury afforded by only the wealthy and sheds a light upon the possibility of going green in Egypt.

Thompson et al. (2010) suggest that the consumers who report the strongest preferences for environmentally certified forest products were more willing to pay a premium and more likely to display environmentally conscious behavior.

Finisterra do Paço et al. (2010) found certain environmental and demographic variables to be significant for differentiating between the "greener" segment and the rest. Tan and Lau (2011) propose an integrated framework to study the role of PCE on the V-A-B model in future green buying behavioral studies. Baqer's (2012) comparative study highlights that Kuwait and Turkey are in serious need of environmental education for their populations to understand the dangers of their pattern of consumption, as compared to the United States. Hofmeister et al. (2012) differentiated four consumer segments on the basis of the NEP scale in Hungarian society.

Athar et al. (2011) examined Pakistani consumers' intentions to buy environmentally friendly products and concluded that consumers are ready to buy green products more frequently if the green products' performance is as competitive as their traditional counterparts. Majláth (2010) verified that green consumers feel that they contribute to both the formation of environmental problems and the solution to these problems through their actions. Arttachariya (2012) found environmental consciousness, concern for environment, and reference group influence as significant predictors of green purchasing behavior, but failed to show any significant relationship in the demographic variable of graduate students in Bangkok.

Siringi (2012) found few factors to be of significance in green consumer behavior. Sharma (2015) provides examples of the firms in India that are going green and creating significant profits and customer satisfaction.

## 14.3 RESEARCH OBJECTIVES

There are multiple research objectives:

- Identify different segments of physicians regarding their environmental sensitivity based on their self-reported attitude, perceived effectiveness, knowledge, and value orientation.

- Validate the segmentation with respect to demographic- and industry-related profiles.
- Attempt to construct different targeting strategies for physicians at different levels of environmental sensitivity.

## 14.4 GREEN SEGMENTATION: A CROSS-SECTIONAL STUDY OF PHYSICIANS OF INDIA

### 14.4.1 RESEARCH METHODOLOGY

This descriptive research is done on a convenient sample of physicians, and judgment has been used when selecting the subjects to ensure that they are from metropolitan areas, cities, and towns; are representative of men and women; and are composed of different age groups, different medical specialities, and different work setups. The pre-tested, structured, and self-administered questionnaire consisted of 25 Likert scale statements taken from established and validated environmental scales, to examine the environmental dimensions. For environmental concerns, NEP is used. This well-accepted 15-item scale has been used in multiple contexts to assess peoples' perceptions concerning the environment throughout the world. Perceived environmental knowledge is measured using the Perceived Knowledge of Environmental Issues scale (Ellen et al. 1991). This 5-statement scale was found to be valid and reliable with a reported $\alpha$ value of 0.86 (Mohr et al. 1998). The PCE variable measures the extent to which a respondent believes that an individual consumer can be effective in pollution abatement. The PCE 5-item scale employed by Choi and Kim (2005) was used in this study to measure those factors. The scale reported to provide an $\alpha$ of 0.74.

A 12-item reliable value scale by Stern and colleagues is used in measuring value orientations (Stern 1999).

A couple of open-ended questions were also incorporated. Out of the 550 questionnaires distributed, 341 were retrieved, and 316 questionnaires were chosen after data cleansing for research purposes. Microsoft Excel and SPSS were used for statistical analysis of data.

#### 14.4.1.1 Reliability and Validity

The outcome shows Spearman Brown Coefficient for unequal length (with 37 items) as 0.933. Here the Cronbach's alpha, inclusive of all scale items, is 0.850. The Kaiser-Meyer-Olkin (KMO) verifies the adequacy of the sample size (KMO = 0.822), and Bartlett's test of sphericity chi-square (105) =1851.060, p = 0.000 revealed that the correlations between the scale items are sufficient for principal component analysis (PCA).

### 14.4.2 DESCRIPTIVE STATISTICS

#### 14.4.2.1 Characteristics of the Dataset

The socio-demographic profile of the dataset, as displayed in Figure 14.1, provides an idea about the distribution and spread of the sample across genders, age groups, employment types, specializations, and regions. Though there is a smaller percentage of women doctors in the community, female doctors responded more frequently

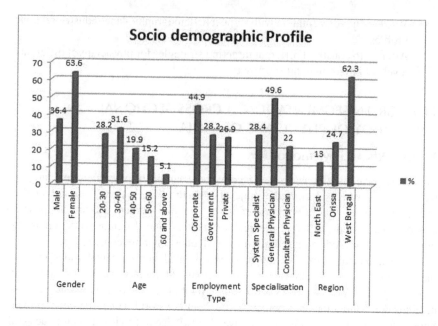

**FIGURE 14.1**   Socio-demographic profile.

because they are more approachable. The age profile reveals that the majority of respondents are younger physicians. Nearly half of the respondents belong to the corporate sector, with government doctors and private practitioners responding in nearly equal numbers. Doctors from different medical specialities have taken part in the survey, and for the ease of calculation and understanding, they are categorized under three headings: system specialists (such as cardiologists, pulmonologists, endocrinologists, dermatologists, anesthetists), general physicians (residential medical officers and family physicians), and consultants in medicine. As expected, general physicians constitute the majority. On the basis of region, more than half of the respondents belong to West Bengal, followed by Orissa and North East.

## 14.4.2.2   Principal Component Analysis

As an extensive number of variables are used in this research, an exploratory factor analysis is used to simplify the interpretation of the variables. A PCA is done and 10 factors are originated. Each of these factors is depicted by means of several items used in the questionnaire. These new factors are used as inputs in the later cluster and discriminant analysis. Varimax rotation technique is employed to improve the components' interpretability.

The KMO test verified the adequacy of the sample size (KMO = 0.856), and Bartlett's test of sphericity chi-square (df, 666) = 6429.526, p-0.000, revealed that the correlations between the scale items were sufficient for PCA. Components with Eigenvalues greater than one are only considered. This 10-component solution explains 72.65% of the total variance. This solution also satisfies the scree test, as seen in Figure 14.2, for extraction of significant factors in factor analysis.

A short description of each component, along with component characteristics, is illustrated in Table 14.1.

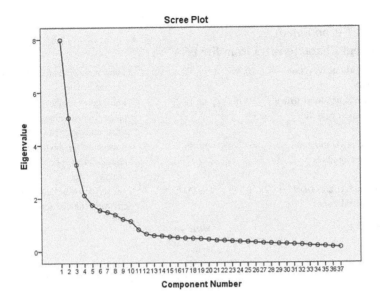

**FIGURE 14.2**   Scree plot.

## TABLE 14.1
## Component Characteristics from the PCA

| *Component 1* | | | *Personal Effectiveness* |
|---|---|---|---|
| | | **Items** | |
| **Personal behavior has positive effects** | 0.777 | Qu. 21 | Each person's behavior can have a positive effect on society by signing a petition in support of promoting the environment. |
| **Belief to be able to solve natural resources problem by conserving water** | 0.805 | Qu. 22 | I feel I can help solve natural resource problem by conserving water |
| **Ability to protect environment by buying eco-friendly products and** | 0.795 | Qu. 23 | I can protect the environment by buying products that are friendly to the environment |
| **Not much one can do about environment** | 0.769 | Qu. 24 | There is not much that I can do about the environment |
| **Capability of solving environmental problems** | 0.799 | Qu. 25 | I feel capable of helping solve the environment problems |
| *Component 2* | | | *Knowledge Perception* |
| | | **Items** | |
| **Buy environmentally safe products** | 0.727 | Qu. 16 | I know that I buy products and packages that are environmentally safe |

*(Continued)*

**TABLE 14.1 (*Continued*)**
## Component Characteristics from the PCA

| | | | |
|---|---|---|---|
| **Know more about recycling than others** | 0.769 | Qu. 17 | I know more about recycling than the average person |
| **Know to select products which reduce waste for landfill** | 0.741 | Qu. 18 | I know how to select products and packages that reduce the amount of waste ending up in landfills |
| **Understand environmental phrases and symbols** | 0.776 | Qu. 19 | I understand the environmental phrases and symbols on product package |
| **Very knowledgeable about environmental issues** | 0.773 | Qu. 20 | I am very knowledgeable about environmental issues |
| *Component 3* | | *Altruism* | |
| | | **Item** | |
| **EQ** | 0.828 | Equality | Equal opportunity for all |
| **WAP** | 0.837 | A world at peace | Free of war and conflict |
| **SJ** | 0.804 | Social justice | Correcting injustice, care for the weak |
| **HLP** | 0.788 | Helpful | Working for the welfare of others |
| *Component 4* | | *Egocentrism* | |
| | | **Item** | |
| **SP** | 0.841 | Social power | Control over others, dominance |
| **WL** | 0.854 | Wealth | Material possessions, money |
| **AU** | 0.839 | Authority | The right to lead or command |
| **INF** | 0.915 | Influential | Having an impact on people and events |
| *Component 5* | | *Biospheric Value* | |
| | | **Item** | |
| **PP** | 0.678 | Preventing pollution | Protecting natural resources |
| **RE** | 0.678 | Respecting the earth | Harmony with other species |
| **UN** | 0.721 | Unity with nature | Fitting into nature |
| **PE** | 0.667 | Protecting the environment | Preserving nature |
| *Component 6* | | *Human Dominance* | |
| | | **Item** | |
| **Human ingenuity is sufficient (R)** | 0.780 | Qu.4 | NEP 4 |
| **Humans are subject to laws of nature** | 0.818 | Qu. 9 | NEP 9 |
| **Humans will control nature (R)** | 0.872 | Qu. 14 | NEP 14 |

**TABLE 14.1** (*Continued*)
**Component Characteristics from the PCA**

| Component 7 | | Limited Resources | |
|---|---|---|---|
| | | Item | |
| Approaching limit of earth | 0.776 | Qu. 1 | NEP 1 |
| Earth has plenty of resources (R) | 0.825 | Qu. 6 | NEP 6 |
| Earth is like a spaceship | 0.809 | Qu. 11 | NEP 11 |
| Component 8 | | Equilibrium of Nature | |
| | | Item | |
| Human interference is disastrous | 0.814 | Qu. 3 | NEP 3 |
| Balance of nature is strong (R) | 0.852 | Qu. 8 | NEP 8 |
| Balance of nature is delicate | 0.759 | Qu. 13 | NEP 13 |
| Component 9 | | Major Catastrophe | |
| | | Item | |
| Humans abuse the environment | 0.782 | Qu. 5 | NEP 5 |
| Eco-crisis is exaggeration (R) | 0.794 | Qu. 10 | NEP 10 |
| Major catastrophe is near | 0.781 | Qu. 15 | NEP 15 |
| Component 10 | | Equal Rights | |
| | | Item | |
| Humans have right to modify nature (R) | 0.840 | Qu. 2 | NEP 2 |
| Plants and animals have equal rights | 0.821 | Qu. 7 | NEP 7 |
| Humans are meant to rule (R) | 0.739 | Qu. 12 | NEP 12 |

### 14.4.2.3 Cluster Analysis

An initial clustering solution is obtained using Ward's method. The dendrogram pictured in Figure 14.3 helps to opt for a five-cluster solution, which is used as input for k means cluster analysis to optimize partitioning.

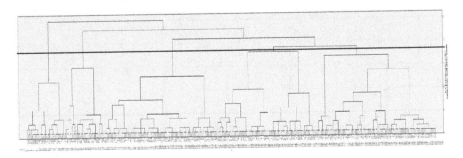

**FIGURE 14.3** Dendrogram using Ward linkage.

**TABLE 14.2**

**Cluster Analysis Report**

|  | Cluster Centroid | | | | |
|---|---|---|---|---|---|
|  | Cluster I | Cluster II | Cluster III | Cluster IV | Cluster V |
| **Clustering Variables** | **n = 42**<br>**(13%)** | **n = 60**<br>**(19%)** | **n = 70**<br>**(22%)** | **n = 44**<br>**(14%)** | **n = 100**<br>**(32%)** |
| Personal Effectiveness | −1.1534 | .1616 | .3563 | .6208 | −.1352 |
| Knowledge Perception | −.1021 | .3145 | .2587 | −1.0688 | .1433 |
| Altruism | .3101 | .1405 | .0575 | −1.0548 | .2092 |
| Egocentrism | .3990 | −.2043 | .5777 | −.5240 | −.2188 |
| Biospheric | .8195 | −.1569 | .2708 | .6303 | −.7169 |
| Human Dominance | .5352 | .2902 | −.6505 | .1392 | −.0048 |
| Limited Resources | .0232 | .5214 | .5158 | −.1742 | −.6071 |
| Equilibrium of Nature | −.7558 | .5114 | .3799 | −.2118 | −.1621 |
| Major Catastrophe | −.0959 | −1.1884 | .4936 | .1351 | .3483 |
| Equal Rights | −.0262 | −.2233 | .1092 | .3231 | −.0736 |

In the second step, the standardized factors are considered as new variables and used in cluster analysis, which includes a combination of a hierarchical and a non-hierarchical approach to divide the sample into segments. The cluster analysis report is given in Table 14.2.

The variables showing the highest positive or negative scores in absolute terms help to explain the inclusion of consumers in the different groups or clusters, as seen in Tables 14.2 and 14.4.

After the optimum numbers of clusters were obtained, several tests were carried out in order to discover whether there were any significant differences among the groups using ANOVA and discriminant analyses.

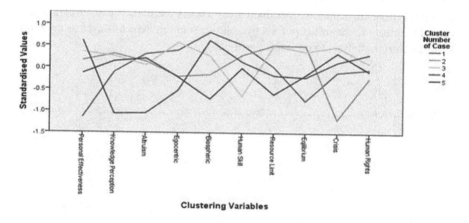

**FIGURE 14.4** Profile diagram.

## TABLE 14.3
## Tests of Equality of Group Means

|  | Wilk's Lamda | F | df1, df2 | Sig. |
|---|---|---|---|---|
| Personal Effectiveness | .730 | 28.793 | 4, 311 | .000 |
| Knowledge Perception | .799 | 19.584 | 4, 311 | .000 |
| Altruism | .813 | 17.841 | 4, 311 | .000 |
| Egocentrism | .843 | 14.471 | 4, 311 | .000 |
| Biospheric | .671 | 38.166 | 4, 311 | .000 |
| Human Dominance | .849 | 13.831 | 4, 311 | .000 |
| Limited Resources | .768 | 23.523 | 4, 311 | .000 |
| Equilibrium of Nature | .827 | 16.228 | 4, 311 | .000 |
| Major Catastrophe | .635 | 44.780 | 4, 311 | .000 |
| Equal Rights | .971 | 2.285 | 4, 311 | .060 |

By observing the F value in Table 14.3, it can be seen that the groups displayed different means. When all the variables were considered individually, except for the "equal rights" variable, they were significant for differentiating between the groups. The obtained results suggest that the variable "major catastrophe" provides the greatest difference between the means of the clusters, as it presents the lowest score. After this, and in descending order of their discriminatory power, come the variables biospheric value, personal effectiveness, limited resource, knowledge perception, equilibrium of natural, altruism, egocentrism, and human dominance, followed by equal rights as a non-significant factor.

The discriminant analysis generated 4 (number of groups — 1) discriminant functions, of which the first two are having eigenvalues higher than one and are statistically significant. So, the remaining two functions are not considered (Table 14.4).

In view of the statistical significance difference observed among the groups, it is useful to examine the individual contribution of the variables to the discriminant functions. The relative contribution of each of the variables to the discriminant function can be analyzed through the structure matrix.

The results were interpreted by examining the standardized discriminant function coefficients, the structural correlations, and all group scatter diagram.

## TABLE 14.4
## Discriminant Function

| Function | Eigenvalue | Cannonical Correlation | Wilk's Lambda | Chi Square | Df | Sig. |
|---|---|---|---|---|---|---|
| $f_1$ | 1.474 | 0.772 | 0.050 | 922.157 | 40 | 0.000 |
| $f_2$ | 1.302 | 0.752 | 0.123 | 643.651 | 27 | 0.000 |

**TABLE 14.5**

**Standardized Canonical Discriminant Function Coefficient**

| Discriminant Function Coefficients | Standardized Canonical | |
|---|---|---|
| | $f_1$ | $f_2$ |
| Personal Effectiveness | −0.254 | 0.537 |
| Knowledge Perception | 0.770 | −0.023 |
| Altruism | 0.690 | −0.276 |
| Egocentrism | 0.180 | 0.125 |
| Biospheric | −0.515 | 0.441 |
| Human Dominance | 0.012 | −0.172 |
| Limited Resources | 0.221 | 0.762 |
| Equilibrium of Nature | 0.343 | 0.552 |
| Major Catastrophe | −0.447 | −0.465 |
| Equal Rights | −0.315 | 0.054 |

The standardized coefficients (Table 14.5) indicate the largest coefficient for knowledge perception on function 1, whereas function 2 has relatively larger coefficients for limited resource and equilibrium of nature.

Figure 14.5 is a scatter diagram plot of all groups on function 1 and 2. It can be seen that group 2 has the highest value on function 1, and group 4 is the lowest. Because function 1 is primarily associated with perception of knowledge, it is expected that the groups would be ordered on that variable. The interpretation is further strengthened by group means regarding knowledge perception.

The plot also indicates the function 2 tends to separate group 2 (highest value) and group 5 (group 5). This function is primarily associated with finiteness of resources and the delicate balance of nature.

Given the high correlation of these variables with function 2 in the structure matrix, group 2 is expected to show more concern toward the limit and balance aspect of our planet. The group means of those groups on these two variables reconfirm the same findings.

To test whether the functions were valid predictors of cluster membership, the classification matrices of the respondents were examined. 96.8% of the 316 respondents were correctly classified into their groups by the discriminant function, as displayed in Table 14.6. The high rate of accuracy was indicative of the reliability of the clusters.

## 14.5 MAJOR FINDINGS: SEGMENTATION OF THE PHYSICIANS WITH DEMOGRAPHIC AND INDUSTRY-RELATED PROFILES

The five clusters were interpreted in terms of the variables and named accordingly: Lip Service Critics, Confused Concerned, Egoist Concerned, Naive Concerned, and Indifferent. The characteristics of the various segments of physicians found through the various statistical procedures are described later. Each cluster has been examined

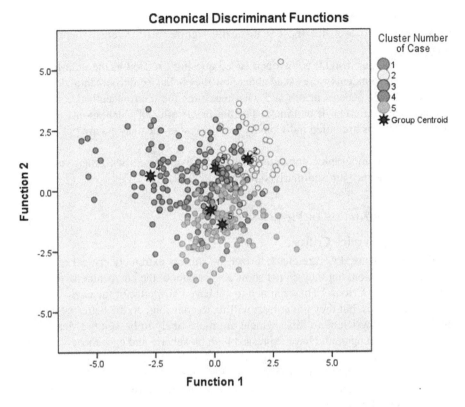

**FIGURE 14.5**   All group scattergram.

## TABLE 14.6
## Classification Matrix

| Classification Matrix | | 1 | 2 | 3 | 4 | 5 | |
|---|---|---|---|---|---|---|---|
| Original count % | 1 | 36 | 0 | 2 | 0 | 4 | 42 |
| | 2 | 0 | 60 | 0 | 0 | 0 | 60 |
| | 3 | 0 | 0 | 69 | 0 | 1 | 70 |
| | 4 | 0 | 0 | 0 | 41 | 3 | 44 |
| | 5 | 0 | 0 | 0 | 0 | 100 | 100 |
| | 1 | 85.7 | .0 | 4.8 | .0 | 9.5 | 100.0 |
| | 2 | .0 | 100.0 | .0 | .0 | .0 | 100.0 |
| | 3 | .0 | .0 | 98.6 | .0 | 1.4 | 100.0 |
| | 4 | .0 | .0 | .0 | 93.2 | 6.8 | 100.0 |
| | 5 | .0 | .0 | .0 | .0 | 100.0 | 100.0 |

*96.8% of original grouped cases correctly classified.

based on their socio-demographic and other environmental variables.. Chi-square tests are used to look for those statistically significant differences between the segments.

To gain deeper insight, a few open ended questions related to the industry were asked. The respondents were asked about how they believe environmental deterioration is affecting general health and what measures the pharmaceutical companies can take to ensure eco-friendliness. For the practicality of calculations, the most common answers are noted individually and the rest of the answers are put together as others.

The socio-demographic characteristics of the five clusters with their respective level of significance are summarized in Table 14.7.

### 14.5.1 Segmentation of Physicians

#### 14.5.1.1 Lip Service Critics

Of the respondents, 13% are characterized by stances extremely critical of human attitudes and action, but they do not show concern about the environment; these are the "Lip Service Critics." They claimed to not have knowledge of the environmental issues themselves, but they are at least willing to contribute to the betterment of the environment. Physicians in this segment are more likely to be young (Mean-2.19), found in West Bengal, and have expressed high biospheric and egocentric value orientation together. These fault-finding individuals claim to care about nature without doing much for it. Their choice of measurement for green pharmaceutical companies is recycling, and most of them came up with no concrete answer when asked about how health is affected by the environment.

The environmental concern of this respondent group is similar to ROPER I's (1990) "Grousers" and Said's (1997) "Lip Service Green."

#### 14.5.1.2 Confused Concerned

Around 19% of the respondents claim to be concerned about the finiteness of resources and delicate balance of Earth systems, but are still confused about the possibility of any ecological catastrophe. Most likely residents of West Bengal, these respondents are not very confident in the belief that they can effectively do something to save nature, and they also claim to have knowledge about environmental issues. These mostly younger (mean = 2.38) altruistic respondents believe in the equal rights of all species and are critical of human attitudes and actions. When asked about the adverse effects the environmental hazard has on people's health, their responses are mostly split between selecting respiratory trouble and malignancy. Their chosen measure for pharmaceutical companies combating pollution is planting trees.

This group exhibits similarities with ROPER I's (1990) "Sprout" cluster.

#### 14.5.1.3 Egoist Concerned

The respondents in "Egoist Concerned" (22%) are most likely to work in corporate or private setups. These physicians usually are either very young or belong to the

# TABLE 14.7
## Physicians' Characteristics between Clusters

| | | Lip Service Critics | Confused Concerned | Egoist Concerned | Naive Concerned | Indifferent | Total |
|---|---|---|---|---|---|---|---|
| **Age** | | | | | | | |
| $\chi^2 = 31.351$ | <30 | 13 | 19 | 21 | 10 | 26 | 89 |
| df = 16 | 31–40 | 18 | 16 | 22 | 15 | 29 | 100 |
| $p < 0.647..$ | | | | | | | |
| | 41–50 | 4 | 12 | 13 | 8 | 26 | 63 |
| | 51–60 | 4 | 9 | 13 | 7 | 15 | 48 |
| | >60 | 3 | 4 | 1 | 4 | 4 | 16 |
| **Gender** | | | | | | | |
| $\chi^2 = 6.267$ | Female | 28 | 34 | 39 | 33 | 67 | 201 |
| df = 4 | Male | 14 | 26 | 31 | 11 | 33 | 115 |
| $p < 0.180..$ | | | | | | | |
| **Employment Type** | | | | | | | |
| $\chi^2 = 21.578$ | Corporate | 21 | 17 | 36 | 23 | 45 | 142 |
| df = 8 | Govt. | 11 | 28 | 12 | 15 | 23 | 89 |
| $p < 0.006..$ | Private | 10 | 15 | 22 | 6 | 32 | 85 |
| **Specialization** | | | | | | | |
| $\chi^2 = 4.888$ | System specialist | 8 | 18 | 21 | 12 | 31 | 90 |
| df = 8 | General physician | 21 | 28 | 37 | 21 | 50 | 157 |
| $p < 0.769..$ | Consultant physician | 13 | 14 | 12 | 11 | 19 | 69 |

(Continued)

**TABLE 14.7** (*Continued*)
**Physicians' Characteristics between Clusters**

| | Lip Service Critics | Confused Concerned | Egoist Concerned | Naive Concerned | Indifferent | Total |
|---|---|---|---|---|---|---|
| **States** | | | | | | |
| $\chi^2 = 13.071$ | | | | | | |
| df = 8 | | | | | | |
| p < 0.109.. | | | | | | |
| Assam | 6 | 7 | 8 | 6 | 14 | 41 |
| Orissa | 6 | 11 | 17 | 19 | 25 | 78 |
| WB | 30 | 42 | 45 | 19 | 61 | 197 |
| **WB & the Rest** | | | | | | |
| $\chi^2 = 10.046$ | | | | | | |
| df = 4 | | | | | | |
| p < 0.040 | | | | | | |
| Non-WB | 12 | 18 | 25 | 25 | 39 | 119 |
| WB | 30 | 42 | 45 | 19 | 61 | 197 |
| **Effect on Health** | | | | | | |
| $\chi^2 = 81.161$ | | | | | | |
| df = 12 | | | | | | |
| p < 0.000.. | | | | | | |
| Respiratory infection | 15 | 25 | 31 | 23 | 13 | 107 |
| Health deterioration | 5 | 7 | 8 | 5 | 57 | 82 |
| Malignancy | 12 | 16 | 18 | 8 | 14 | 68 |
| Others | 10 | 12 | 13 | 8 | 16 | 59 |
| **Measure to Be Taken by Pharmaceuticals** | | | | | | |
| $\chi^2 = 57.046$ | | | | | | |
| df = 12 | | | | | | |
| p < 0.000.. | | | | | | |
| Recycling | 12 | 10 | 15 | 4 | 18 | 59 |
| Tree plantation | 5 | 30 | 25 | 22 | 15 | 97 |
| Waste reduction | 9 | 9 | 19 | 8 | 16 | 61 |
| Others | 16 | 11 | 11 | 10 | 51 | 99 |

50–60 age bracket. They show high concern for all environmental issues. They claim to have environmental knowledge, and most of them mention that pharmaceutical companies need to carry out proper waste disposal. Having confidence in their ability to protect nature, members of this segment are not critical of human attitudes and action; rather, they believe humans, with their technological ingenuity, are going to save the earth. The overall picture seems to portray a traditional environmental welfare stance backed by the dominance of human development (Edwards-Jones et al. 2000). Their strong rejection of ecocentric values suggests that this highly egocentric segment of physicians holds a more traditional anthropocentric worldview.

"Egoist Concerned" have a similar kind of disposition to the "Approvers" and "Egocentric Pushers" segments studied by Zografos and Allcroft (2007) and Hofmeister et al. (2012), respectively.

### 14.5.1.4   Naive Concerned

"Naive Concerned" (14%) constitutes a highly eco-crisis-worried group of doctors who are most likely to be on the higher side of the age spectrum (Mean = 2.55) and show less concern for the planetary ecosystem limits and balances. They believe in their ability to positively influence and solve environmental problems. These biospheric value oriented respondents are highly critical of human attitude and believe in equality of all species. Mostly from Orissa, these doctors are less likely to be self-employed. Their indifference toward earth's eco-balance may be the outcome of having less knowledge on issues of the environment. Planting trees is their recommendation to the pharmaceutical companies, and they feel lung diseases are most common due to the deteriorating environment.

### 14.5.1.5   Indifferent

The "Indifferent" segment comprises nearly one-third (32%) of the respondents who show very negative positions in relation to nearly all environmental aspects; they are not willing to take action to solve environmental problems. Being non-critical of prevailing human attitudes, they are convinced that humans should rule the planet. With highly negative biospheric orientation, these are likely to be elderly private practitioners from all states, and they claim to have lesser understanding of nature. They mention different measures, some even irrelevant, for the pharmaceutical companies to take and mostly have mentioned general health deterioration as the effect of environmental degradation. These answers confirm that members of this group do not pay much attention to environmental issues, including in their professional field. In many ways, this group resembles ROPER I's (1990) "Basic Brown" cluster, Said's (1997) "Brown Bombers," and Finisterra do Paço et al. (2009) "Uncommitted."

### 14.5.2   Research Contribution

This academic research contributes to the pharmaceutical industry by providing evidence of the existence of a viable market for drugs with lower environmental impacts and by constructing targeting strategies to attract physicians at different levels of environmental sensitivity in a sound theoretical context. The physicians in eastern India belong to different strata of environmental sensitivity. This study segments the pharmaceutical market of eastern India on the basis of prescribers' demographic, psychographic, knowledge, and value orientation. This study reveals that around 55% of the physicians (combining confused, egoist concerned, and naive concerned) who are concerned about the environment are likely to be receptive to green appeals. This group is large enough to warrant the attention of marketers.

Rather than disturbing their traditional anthropocentric views of humans as rulers of the planet, the Egoist Concerned segment should be encouraged to look after nature as their subject. A green positioning strategy focusing on emotional benefits, such as happiness derived through helping the environment, feeling of well-being by sacrificing time, cost, and effort for environmental causes, may be emphasized (Chan and Yam 1995; Hartmann et al. 2005). As imitator physicians value opinion leaders for prescription guidelines (Nickum 2007), engaging opinion leaders from this segment would be a good marketing ploy. As Ottman (2011) recommends "Engage, enlighten and entertain" strategy for successful green marketing.

The Confused Concerned are passionate about environmental issues but fail to understand the severity of the consequences of the environments' rapid degradation. This comparatively young group is less confident about the part they should play to save the world. They need to be convinced that they actually can take charge and contribute. On the other hand, the Naive Concerned are worried about the possibility of crisis without having much of an understanding about limit and balance of Earth. The Naive Concerned are extremely enthusiastic about their share of accountability. Their accountability has to be encouraged through positive feedback, supported by evidence or providing simple statistics, that their green prescriptions are helping nature. According to Choi and Kim (2005) high self efficacy has direct influence on green purchase behavior. Both the Confused Concerned and the Naive Concerned have a need for knowledge that must be satisfied to achieve the green marketing objectives. The positive biospheric value orientation of these two groups has to be triggered to action.

The Indifferent and the Lip Service Critics are the most likely to be reluctant to respond to environmentally sensitive messages. Both groups have claimed to have less awareness about environmental issues and are not willing to take personal responsibility toward the environment. Becken (2007) stated that information and knowledge are a crucial precursor for pro-environmental actions. For the Lip Service Critics, playing on their fear and guilt appeal may work. Physicians who keep environmental concern in mind when they prescribe may be highlighted as belonging to a desired league, and they can be used to encourage those who do not care about the environment. The Lip Service Critics, being young and exhibitionist in temperament, may be motivated by promoting "green as trendy."

Porter and van der Linde (1995) warned economically developing countries that promote resource-wasting methods that are detrimental to environmental well-being, and perceive green methods as being "too expensive" to employ, will remain uncompetitive in the long run. The Indian pharmaceutical industry, being a probable candidate for Foreign Direct Investment (FDI), cannot ignore the global shift in paradigm toward embracing greener products and methodologies.

The firms which do not respond to the "green challenge" with their product and production facilities will eventually risk credibility in the eyes of environmentally conscious customers. On the contrary, the genuine green firms that exercise green marketing strategies will be able to obtain a competitive advantage by offering this emerging attribute and will be able to get countless opportunities to build a sustainable brand image.

### 14.5.3 LIMITATION AND SCOPE FOR FURTHER RESEARCH

One of the major methodological limitations of this survey is that the collected data represent the respondents' perceptions, which are assumed to be factual and genuine. As the research is on a socially sensitive topic, and the respondents belong to a socially elite group as physicians, distortion, and bias may creep in to maintain social desirability. A pan-India study will improve the generalizability of the current research. Longitudinal studies to monitor the shift in the segment proportion will also be a very intriguing research topic.

## REFERENCES

Arttachariya, Patricia. 2012. "Environmentalism and green purchasing behaviour: A study on graduate students in Bangkok, Thailand." *BU Academic Review* 11(2): 1–11.

Athar, Afzaal Ali., Ali Khan, and Israr Ahmed. 2011. "Determinants of Pakistani consumers' green purchase behaviour: Some insights from a developing country." *International Journal of Business and Social Science* 2(3): 217–226.

Attfield, Robin. 1981. "The good of trees." *Journal of Value Inquiry* 15(1): 35–54.

Baqer, Samar. 2012. "True green consumers: An investigation of consumers' genuine willingness to share environmental responsibility." *Global Journal of Business Research* 6.

Becken, S. 2007. "Tourists' perception of international air travel's impact on the global climate and potential climate change policies." *Journal of Sustainable Tourism* 15: 351–368.

Belz, F. M. 2006. "Marketing in the 21st century." *Business Strategy and the Environment* 15(3): 139–144. http://dx.doi.org/10.1002/bse.529.

Berger, I. E., and R. M. Z. Corbin. 1992. "Perceived consumer effectiveness and faith in others as moderators of environmentally responsible behaviors." *Journal of Public Policy & Marketing* 11: 79–89.

Cambridge Reports. 1993. "Green consumers- the impact of environmental concern on consumers." Research International Inc.

Cegarra-Navarro, J.-G., J-Rodrigo Cordoba-Pachon, and G. W. Fernandez de Bobadilla. 2009. "Creating environmental knowledge through 'green communities' in the Spanish pharmaceutical industry." *The Service Industries Journal* 29(12): 1745–1761.

Chan, R. Y.-K., and E. Yam. 1995. "Green movement in a newly industrializing area: A survey on the attitudes and behaviour of the Hong Kong citizens." *Journal of Community & Applied Social Psychology* 5(4): 273–284.

Choi, Sejung, and Yeonshin Kim. 2005. "Antecedents of green purchase behavior: An examination of collectivism, environmental concern, and PCE." *Advances in Consumer Research* 32: 592–599.

Dunlap, Riley E., and D. Van Liere Kent. 1978. "The 'New Environmental Paradigm'." *The Journal of Environmental Education* 9:4, 10–19.

Edwards-Jones, G., B. Davies, and S. Hussain. 2000. *Ecological Economics: An Introduction.* Oxford: Blackwell Science.

Ellen, P. S., Joshua Wiener, Cobb Lyle, and Cathy Walgren. 1991. "The Role of Perceived Consumer Effectiveness in motivating Environmentally Conscious Behaviour." *Journal of Public Policy & Marketing* 10(2): 102–117.

Finisterra do Paço, Arminda F., Mario Raposo, and Walter Filho. 2009. "Identifying the green consumer: A segmentation study." *Journal of Targeting, Measurement and Analysis for Marketing* 17: 17–25.

Finisterra do Paço, A. M., and M. L. B. Raposo. 2010. "Green consumer market segmentation: empirical findings from Portugal". *International Journal of Consumer Studies* 34: 429–436.

Fryxell, Gerald E., and W. H. Lo. Carlos. 2003. "The influence of environmental knowledge and values on managerial behaviours on behalf of the environment: An empirical examination of managers in China." *Journal of Business Ethics* 46(1): 45–69.

Ham, Marija. 2009. "Consumer segmentation based on the level of environmental responsibility." *TRŽIŠTE UDK 658.89:502.12* 21(2): 183–202.

Hartmann, P., V. Apaolaza Ibáñez, and F. J. Forcada Sainz. 2005. "Green branding effects on attitude: functional versus emotional positioning strategies." *Marketing Intelligence & Planning* 23(1): 9–29.

Hogg, M., and G. Vaughan. 2005. *Social Psychology* (4th edition). London: Prentice- Hall.

Hofmeister, Ágnes, Kata Kasza-Kelemen, and Marianna Piskóti. 2012. "Life paths in Hungary in the light of commitment to sustainability." *Interdisciplinary Environmental Review* 13: 323–339. doi: 10.1504/IER.2012.051449.

Jain, Sanjay K., and Gurmeet Kaur. 2004. "Green marketing: An attitudinal and behavioural analysis of Indian consumers." *Global Business Review* 5(2): 187–202.

Kardash, W. J. 1974. "Corporate responsibility and the quality of life: Developing the ecologically concerned consumer." In *Ecological Marketing*, eds. K. E. Henion and Kinner. *Towards Sustainability: The Third Age of Green Marketing.* https://www.researchgate.net/publication/233619700Towards_Sustainability_The_Third_Age_of_Green_Marketing.

Kinnear, T., J. Taylor, and S. Ahmed. 1974. "Ecologically concerned consumers: Who are they?" *Journal of Marketing* 38(1): 20–24.

Kotchen, Matthew, and Stephen D. Reiling. 2000. "Environmental attitudes, motivations, and contingent valuation of nonuse values: a case study involving endangered species." *Ecological Economics* 32(1): 93–107.

Larasati, Yohana, and Arga Hananto. 2012. "The role of value congruity and consumer brand identification toward development of brand commitment and positive word of mouth." *SSRN Electronic Journal.* https://ssrn.com/abstract=2200313.

Laroche, M., and J. Bergeron. 2001. "Targeting consumers who are willing to pay more for environmentally friendly products." *Journal of Consumer Marketing* 18(6): 503–520.

Majláth, Melinda. 2010. "Can Individuals do anything for the environment? – The role of perceived consumer effectiveness' *Proceedings of FIKUSZ." 10 Symposium for Young Researchers* 157–166.

Menon, M. 2014. "Indian companies decouple growth from carbon emissions: CDP report." *The Hindu.* https://www.thehindu.com/business/Industry/Indian-companies-decouple-growth-from-carbon-emissions-CDP-report/article11075611.ece

Menon, A., A. Menon, J. Chowdhury, and J. Jankovich. 1999. "Evolving paradigm for environmental sensitivity in marketing programs: a synthesis of theory and practice." *Journal of Marketing Theory and Practice* 7: 1–15.

Menon, Anil, and Ajay Menon. 1997. "Enviropreneurial marketing strategy: The emergence of corporate environmentalism as market strategy." *Journal of Marketing* 61(1): 51–67.

Mohani, V. 2014. "5 Indian companies in global A list of green firms." *The Times of India.* Accessed November 11, 2019. http://timesofindia.indiatimes.com/home/environment/global-warming/5-Indian-companies-in-global-A-list-of-greenfirms/articleshow/44831606.cms.

Mohr, L. A., D. Eroglu, and P. Ellen. 1998. "The development and testing of a measure of skepticism toward environmental claims in marketers' communications." *Journal of Consumer Affairs* 32(1): 30–55. *JSTOR*, www.jstor.org/stable/23859544. Accessed 30 Oct. 2020.

Mostafa, Mohamed M. 2007. "Gender differences in Egyptian consumers' green purchase behaviour: the effects of environmental knowledge, concern and attitude." *International Journal of Consumer Studies* 31: 220–229.

Naess, A. 2003. "The deep ecology movement: Some philosophical aspects." *Environmental Ethics: An Anthology* 262–274.

Nickum, Chris. 2007. "Optimal sales force efforts in specialty care markets: New, more precise measures increase sales force effectiveness." *IMS Health*. [Cross ref: https://www.jstor.org/stable/40472171?seq=1#page_scan_tab_contents].

Ottman, J. 2011. *The New Rules of Green Marketing: Strategies, Tools, and Inspiration for Sustainable Branding*. Berrett-Koehler, San Francisco.

Polonsky, M. J. 1994. "An introduction to green marketing." *Electronic Green Journal* 1(2). Accessed on January 17, 2009. https://escholarship.org/uc/item/49n325b7.

Poortinga, W., L. Steg, and C. Vlek. 2004. "Values, environmental concern, and environmental behaviour: a study into household energy use." *Environment and Behaviour* 36(1): 70–93.

Porter, M. E., and Claas van der Linde. 1995. "Green and competitive: Ending the stalemate." *Harvard Business Review* 73(5) (September–October): 120–134.

Roper Organisation. 1990. "The environment: Public attitudes and private behaviour." Roper *I*. S. C. Johnson Inc.

Roper Organisation. 1993. "The environment: Public attitudes and private behaviour." Roper *II*. S. C. Johnson Inc.

Said, D. 1997. *Green Australia: Mapping the Market Survey*. St. Leonards: Prospect Publishing.

Saxena, R. P., and P. K. Khandelwal. 2008. "Consumer attitude towards green marketing: an exploratory study." *European Conference for Academic Disciplines*. research-pubs@uow.edu.au.

Schwartz, S. H. 1992. "Universals in the content and structure of values: Theoretical advances and empirical tests in 20 countries." *Advances in Experimental Social Psychology* 25: 1–65.

Schwartz, S. H. 1994. "Are there universal aspects in the content and structure of values?" *Journal of Social Issues* 50: 19–45.

Sharma, Neeraj K. 2015. "Industry initiatives for green marketing in India." *Business and Economics Journal* 7(1).

Shrikanth, R., and D. S. N. Raju. 2012. "Contemporary green marketing: Brief reference to Indian scenario." *International Journal of Social Science and Interdisciplinary Research* 1(1). https://www.researchgate.net/publication/264887494Contemporary_green_marketingbrief_reference_to_Indian_scenario.

Siringi, Ranjit. K. 2012. "Determinants of green consumer behaviour of postgraduate teachers." *IOSR Journal of Business and Management* 6(3): 19–25.

Stern, P. C. 1999. "Information, incentives, and pro environmental consumer behaviour." *Journal of Consumer Policy* 22: 461–478.

Tan, Booi-Chen, and Teck-Chai Lau. 2011. "Green purchase behaviour: Examining the influence of green environmental attitude, perceived consumer effectiveness and specific green purchase attitude." *Australian Journal of Basic and Applied Sciences* 5(8): 559–567.

Tantawi, P., N. Shaughnessy, K. Gad, and M. A. S. Ragheb. 2009. "Green consciousness of consumers in a developing country: A study of Egyptian consumers." *Contemporary Management Research* 5(1): 29–50.

Tesser, A., and D. R. Shaffer. 1990. "Attitudes and attitude change." *Annual Review of Psychology* 41: 479–523.

Thompson, Derek W., Roy C. Anderson, Eric N. Hansen, and Lynn R. Kahle. 2010. "Green segmentation and environmental certification: Insights from forest products." *Business Strategy and the Environment* 19: 319–334.

Zografos, Christos, and David Allcroft. 2007. "The environmental values of potential eco-tourists: A segmentation study." *Journal of Sustainable Tourism* 15(1): pp. 44–66.

# 15 Developing Cleaning and Building Management Strategies for Green and Sustainable Healthcare Facilities

*Emon Chakraborty*

## CONTENTS

## 15.1  INTRODUCTION

Environmental costs are significant at the downstream end of healthcare systems and pose a potential threat to the environment. Though most medical waste is regulated to some extent, particularly in the case of infectious materials and biohazardous agents, the treatment of the wastes generated at the upstream end of the healthcare system have yet to be established. Sustainability-based practices in the healthcare system can be incorporated into hospitals and its related environment where the customers and employees play a vital role in the functioning of the healthcare systems. This chapter details the main parameters on which the healthcare industry incorporates sustainability into their businesses. Additionally, this chapter will review and discuss the green health strategy benefits that generally operate from a smaller local range to a much wider global range. Introducing green cleaning is a stepping stone toward achieving green healthcare. Furthermore, this chapter considers the importance of green buildings related to the principles of building information modeling (BIM) and energy modeling. A key factor in determining environmental performance is energy, especially in healthcare facilities because they use almost ten times more energy than office buildings. Finally, energy modeling strategy in green healthcare is discussed.

## 15.2  METHODOLOGY

### 15.2.1  Determination of Factors Influencing Sustainability in Healthcare

Several parameters that influence the incorporation of sustainability into healthcare have been studied. The high increase in the cost of healthcare in most countries has urged many organizations to focus on sustainability as an option to reduce operating costs. With governmental bodies becoming increasingly concerned, they are raising awareness within society on the proper use and availability of natural resources. Their efforts have placed a great demand on the healthcare system to apply sustainable practices to its operations for environmental protection. In the process of

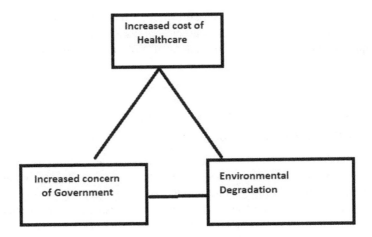

**FIGURE 15.1**   Factors influencing sustainability in healthcare.

implementing sustainable practices, healthcare providers too have realized the significance of simultaneously offering high quality services to patients and maintaining a proper balance between available resources and patient needs.

Furthermore, the increased rate of environmental degradation, mostly through global warming, pollution, and shortage of natural resources (energy and water), serves as a vital driving force for the healthcare industry to explore ways for incorporating sustainability-based practices in its operation. Figure 15.1 schematically represents the various interrelated factors that have an impact on sustainability in healthcare sectors. These circumstances have governments and non-governmental organizations (NGOs) integrating sustainability into healthcare through formulating policies and implementing rules with an objective to save energy and an aim to protect and preserve the environment.

### 15.2.2   EVALUATION OF FRAMEWORK GUIDING TOWARD GREEN HEALTHCARE

The guiding framework for the strategies used to achieve green healthcare expands from a small local scale to a large global scale. On a local scale, a hospital, research facility, or clinic that uses green construction and operation will protect patients, their caregivers, and the patients' families. The healthcare industry should work toward reducing a hospital's ecological footprint in the broader community as well as a larger global range.

Environmental cleaning plays a pivotal role in preventing several health associated infections (HAIs). The method of green cleaning is a novel approach that aims to benefit human health by reducing environmental damage while maintaining or improving the hygiene of the healthcare environment for the purpose of preventing infection. Guidelines established by scientifically oriented third party organizations, such as Green Seal and Ecologo, are used to classify eco-friendly green cleaning products used in the cleaning industry and healthcare sectors (Garza et al. 2015).

The systematic approach of green cleaning not only focuses on the chemical composition of cleaners identified as green, but it also aims to reduce the negative impact of hospital cleaning on human health and the environment (Quan et al. 2011).

### 15.2.3 STUDY OF THE GREEN BUILDING APPROACH THROUGH BIM AND ENERGY MODELING

The most challenging components of healthcare facilities include the planning, designing, construction, and the modes of operation. These facilities are created to maintain and support the most delicate and expensive ventures, such as multiple patient therapy, scientific and diagnostic testing, waste disposal mechanisms involving dry heat and moist heat principles, and food preparation. Healthcare's in-house facilities that have the proper allocation of space for healthcare professionals need extremely sophisticated and large buildings to ensure the hospital functions smoothly with negligible cross contamination.

Across hospital settings, mortality and patient injury due to preventable errors frequently occur. These rates have reportedly affected the emotional well-being of hospital employees, nurses, and doctors. This crisis culminated in the introduction of Lean methodology, which may help eliminate roadblocks to allow a focus on providing care and enhancing the quality of care in the hospitals while reducing avoidable errors (Sarhan et al. 2019).

As a significant consumer of resources and producer of waste, the healthcare industry has a very high ecological footprint. Lean construction, a novel approach for green buildings, primarily focuses on maximizing the value of patient care while minimizing waste. The main principle of Lean technology is based on specifying and identifying patients' values, followed by specific actions that are mindful of patient care and satisfaction. Even small errors during the delivery process can lead to overdoing profligate work, lagging behind, causing changes, and significant overproduction within sustainable buildings. The process waste can undermine the sustainable outcomes, thereby limiting the business model for achieving sustainability. Decreasing process waste in the delivery of green buildings can pave the way for the efficient implementation of sustainable strategies, overcoming the early higher expenses. Efficient time management with judicious material use and effort can be a worthy initiative to provide value for the end users.

As a basic principle of green health, healthcare facilities must be designed with a focus on the health and well-being of patients and staff, along with the clinical outcomes. The BIM serves as a living repository of information, working mostly with facility management to control risks and errors and predicting problems ahead of time to maintain patient safety.

Energy modeling plays an important part in delivering a green building and has compelled healthcare designers to launch programs like DOE 2.1, Ecotect, eQuest, and IES. The main objective of energy modeling is to reduce energy use to improve environmental performance.

## 15.3 RESULTS

### 15.3.1 IMPLEMENTATION OF FACTORS TO IMPROVE SUSTAINABILITY IN HEALTHCARE

#### 15.3.1.1 Maintenance of Social Responsibility and Organizational Culture

Maintaining an organizational culture, social responsibility, ethical behavior, and the drive toward innovation have helped incorporate sustainability into healthcare organizations. Many healthcare organizations, such as the World Health Organization (WHO), UNAIDS, International Telecommunication Union (ITU), World Bank, and UNESCO, have adopted the initiative to develop health information, or e-health strategy, with an objective to provide patients with good treatment and services. Healthcare facilities implementing sustainability-based operations have also paved the way for the enhanced development of activities like staff participation on steering committees and staff learning assessments. As a social responsibility, office employees in hospitals should take the initiative to clean their work spaces so that the cleaning staff are not overburdened.

#### 15.3.1.2 Implementation of Patient-Friendly Approach in Healthcare

In hospitals, one of the vital measures in healthcare livability is the level of service offered to patients through patient care and affordable medical bills, leading to patient contentment. Information technology plays a pivotal role in directing health organizations toward sustainability by increasing efficiency, creating tools for sustainability, and redesigning products into services that make the healthcare system patient-friendly, for instance by allowing medical staff to send vaccine or appointment reminders to patients well ahead of time.

#### 15.3.1.3 Execution of Psychological Sustainability

In a psychological perspective, sustainability takes into consideration the well-being of all people, along with the ecological and social environment. The newly developed approach of sustainability emphasizes promotion rather than circumvention, and this approach favors enrichment, growth, and flexible changes over exploitation, depletion, and irreversible alteration. Sustainability in a hospital can be improved by conducting community-based psychological research among medical practitioners to understand the behaviors, attitudes, and culture. The increased collaboration between different specialities, including nonmedical professionals, would also facilitate the healthcare organization to overcome sustainability challenges.

#### 15.3.1.4 Implementation of Sustainability in Terms of Resource Protection

Environment-oriented sustainability aims to control environmental pollution by reducing waste in operations (e.g. waste treatment and reuse, water recycling, and minimizing the use of hazardous chemicals), along with a goal to help save operating costs. Combining these baseline practices together with the objective of recycling can serve as a vital step to reduce rising quantities and treatment costs and can be used to achieve environmental sustainability in healthcare.

230

### 15.3.1.5 Role of Government and NGOs

The dynamism shown by the healthcare system in collaborating with relevant associations and NGOs is a remarkable step toward incorporating sustainability in healthcare. The internationally recognized prime resources in the healthcare sustainability movement include the United States Green Building Council (www.usgbc.org), Green Guide for Healthcare (www.gghc.com), Practice Green Health (http://practicegreenhealth.org/), American Society for Healthcare Engineering (www.ashe.org), Global Health and Safety Initiative (www.ghsi.ca), and Healthcare Without Harm (http://noharm.org). Health Care Without Harm, a collaboration between activists and healthcare organizations, recommends eliminating mercury from healthcare products. Separating polyvinyl chlorine (PVC) plastics from infectious wastes was proposed because it was observed that when infectious waste was incinerated, the presence of PVC products results in a release of carcinogenic dioxins. The recent coalition considering the safety of the patients and healthcare professionals protested against the use of vinyl intravenous bags containing phthalate plasticizers, which may leach toxins into patients' bodies (Garza et al. 2015). Figure 15.2 depicts the various factors that can be implemented for improvement of sustainability in healthcare.

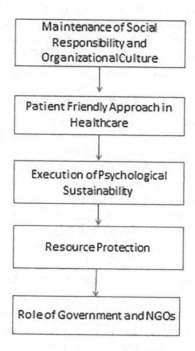

**FIGURE 15.2** Implementation of factors for improving sustainability in healthcare.

## 15.3.2   Functioning of the Framework for Mode of Operations toward Green Healthcare

### 15.3.2.1   Protection of Resources to Reduce the Ecological Footprint

The functional operational framework for attaining green healthcare has the purpose of protecting scarce resources and reducing the rate of global environmental degradation. These goals can be achieved by establishing appropriate sanitation levels in different healthcare settings based on a careful examination of infection risk. This justifies restricted consumption of different valuable resources. For example, the overuse of strong chemicals should be avoided for cleaning administrative offices where infection transmission may be relatively low. A hospital's ecological footprint may also be reduced by efficiently connecting a hospital to its community through pedestrian infrastructure and mass transit. This may prove beneficial in reducing motor vehicle traffic, resulting in better and cleaner air quality in the vicinity of the healthcare facility and community. The aim to protect major renewable and nonrenewable resources can also reduce the ecological footprint.

### 15.3.2.2   Implementation of Green Cleaning Practices

Third party organizations like Green Seal and Ecologo play a pivotal role in green cleaning with a vision to ensure a safer, healthier building, thereby providing better indoor air quality and a lower environmental footprint (Garza et al. 2015). The choice of the cleaning products and procedures is mostly based on different areas' vulnerability to infection cross-transmission.

### 15.3.2.3   Chemical Properties of Green Cleaners

The guidelines for potential "Biologically Based Cleaning and Degreasing Compounds" are mostly recommended while designing a certified green cleaner (Markkanen et al. 2009). A green cleaner should not be toxic nor have any intentionally added carcinogens, mutagens, or reproductive toxins that possibly cause harm to humans and the environment. Preferred cleaning products should use the fewest harmful chemicals. Additionally, the product must have very limited effects on aquatic life and should have the potential to be aerobically biodegraded. It was further mentioned in the guidelines that the certified product should contain less than 1% volatile organic compounds (VOC) at a ready-to-use concentration and should have a pH range between 3.0 and 11.0. It is preferable for the product to be in a concentrated format and be formulated with limited use of natural resources. The formulated product must be effective in preventing various HAI, have the possibility of being recycled, and of reducing waste generation.

### 15.3.2.4   Green Cleaning Components

The different green cleaning components include carpet cleaner, glass cleaner, grease removers, and organic cleaner with citric acid and peroxyl components that is used on all common surfaces that do not need to be disinfected. A light duty green cleaner is used on floors except for operating rooms, which should always use a disinfectant. A common light duty disinfecting agent can be used to disinfect surfaces patients frequently touch. Sensitivity and allergy concerns drove the shift from the old style floor finishes with heavy metals and ammonia to green finishes (Quan et al. 2011).

The different cleaning tools or equipment that use less water and chemicals are sought-after for green cleaning techniques. These products mainly involve light microfiber mops and cloths that help reduce the amount of water and chemicals used, generating less waste than the traditional mops and cloths. Using only one mop head per patient room lowers the risk of cross contamination. A floor scrubbing machine with specialized floor pads and a lower volume of floor stripper can also be introduced to more effectively remove building waste from large floor areas. Using spray bottles with electrically charged water is further recommended for cleaning surfaces without chemicals.

### 15.3.2.5 Dispensing Methods of Green Cleaners to Achieve Sustainable Cleaning

The optimization of application and dispensing methods forms a basis of green cleaning. The wide use of metered chemical dispensers will help reduce spills and chemical waste by carefully dispensing chemicals at the proper dilution rate. The novel approach of using a bottle with a pour spout to place chemicals onto a cleaning cloth will help prevent chemical aerosols from spreading to the indoor air. Directly applying a cleaning solution from a bucket to cleaning aids and cloths through capillary action will ensure the distribution of the exact amount of cleaning solution. Folding a cleaner-saturated mop to create more cleaning surfaces justifies using fewer cloths to clean the same amount of space. Using a new floor finish applicator will prevent odors and VOCs from escaping into the air and can be used to contain floor finish in a box during application (Quan et al. 2011).

The innovative use of fluorescent markers will be helpful in evaluating and improving the thoroughness of cleaning by providing accurate, objective feedback to the cleaning staff to aid in training staff. This method can be done by putting fluorescent markers on high touch surfaces before the cleaning, then using a black light to check if the markers were removed after the cleaning. Through using safe disinfectants or limiting the use of pesticides, there is the possibility of reducing people's exposure to potential toxicity.

### 15.3.2.6 Role of Green Cleaning in Environmental Protection

The systematic approach of green cleaning focuses on the chemical composition of green cleaners and challenges the negative impact of hospital cleaning on human health and the environment. Most healthcare facilities focus on selecting floor materials that are easy to clean and maintain. After closely monitoring the sustainability of interior design, the interior design committee mostly prefers materials that are easy to clean (such as vinyl furniture) and flooring (such as rubber flooring and ceramic tiles) (Quan et al. 2011). Using a floor mat system at entrances, vacuuming, and sweeping entrances will help reduce the amount of soil and dirt inside, eliminating the need for chemical cleaning. The need for cleaning can also be lowered by limiting the number of surfaces, optimizing the heating ventilation and air-conditioning (HVAC) system, moving furniture, and conveniently locating housekeeping closets.

## 15.3.3 The Novel Approach of Green Building through BIM and Energy Modeling

Energy modeling and green building technology serve as a new approach for achieving green and sustainable healthcare.

### 15.3.3.1 Role of Building Information Modeling in Green Buildings

Building information modeling (BIM) is the process of generating and managing building data during its life cycle (Arayici and Aouad 2010). This mode of operation serves as an important component in improving patient safety and quality of patient care, along with being an initiative to provide a safer and better work environment for the staff.

For healthcare facilities, BIM can provide details of tedious functions capable of generating characteristic blueprints. BIM can be used as a model to analyze the commitment necessary to meet a healthcare facility's functional requirements, including expensive medical equipment. Through BIM, the design phase can simulate and study data on the continual glide of individuals and matter that was gathered during the commencement phase. BIM serves as an exclusive data model of the facility that is capable of supporting teamwork and collaboration during the facility's design and construction phases. BIM is further known to assist in integrating architectural design and engineering design for new facilities, as well as for remodeling and incorporating new designs to existing healthcare facilities. A critical component for the healthcare industry's use of BIM is the three-dimensional model of the built facility that can support facility operations and the management team (Arayici and Aouad 2010).

A hospital with complex systems is the sole operating system in which the construction includes ongoing renovations and expansions to accommodate new processes and technologies. BIM is extremely beneficial for the healthcare industry because of the building systems' complexity in terms of indoor environmental quality, cooling and heating loads, and countless medical equipment. Additionally, BIM acts as an active archive of data, aiming to maintain patient safety while controlling risk and process errors (Arayici and Aouad 2010).

### 15.3.3.2 Role of Physical Built Environment in Green Healthcare

There is significant validation that justifies the vital role of ambience in increasing the well-being of patients while providing a convenient environment for the duty staff. Many hospital-acquired infections and patient falls can be avoided if the physical environment is during the planning and building stages of a new construction. The high efficiency particulate air filtration systems (HEPA) can be effective in treating airborne infections by providing the hospital with enhanced ventilation and indoor air quality.

From the patient's perspective, a green building can be advantageous by providing good exposure to daylight, nature, and gardens. These components are capable of reducing anxiety and depression among the patients. Single patient rooms are preferable, as they help lower stress levels from noise. Increased noise levels can contribute to blood pressure elevation, decreased oxygen saturation, increased heart rate, and can also be a major factor for reduced sleep. Single rooms provide a better recovery model by further reducing the chances of infection from cross-contamination.

In the mission of going green, the healthcare facilities will be able to limit the excess energy consumption in their data centers and save money. IT (Information Technology) plays a vital role in assisting healthcare facilities with their green initiatives. Integrating IT operations and virtualizing servers in healthcare can be an efficient method to reduce their carbon footprints. The prime areas of consideration are efficient e-waste management strategies and the judicious use of energy resources. Introducing electronic medical records (EMR) and practicing telemedicine are also strongly supported to achieve green healthcare. These practices can minimize the use of paper and serve as a suitable method for maintaining large patient records for future treatment. Physicians can also refer to EMR records at any time to check on patients' medical records. The concept of telemedicine is being widely practiced during the COVID-19 pandemic to minimize frequent visits to hospitals.

### 15.3.3.3   Role of Energy Modeling toward Attaining Green Healthcare

Healthcare facilities use more energy compared to other facilities. Their environmental performance and sustainability depend on their energy use. Introducing energy system modeling can be an efficient tool to reduce the use of energy. This can be achieved by implementing energy modeling for maintenance of indoor air temperature, floor, and wall designing. The amalgamation of building planning and galvanization with the proper HVAC systems serve as the most accepted method to reduce waste with judicious energy use, thereby limiting harsh environmental consequences. The design of green buildings is largely dependent on energy modeling strategies. Health management officials have listed various programs like DOE 2.1, Ecotect, eQuest, and IES as an energy modeling venture (Oduyemi and Okoroh 2016).

### 15.3.3.4   DOE 2.1

DOE 2.1 can be a useful tool for analyzing energy use per hour and predicting a building's energy cost. The prerequisites include an hourly weather update, a construction design with energy loads description, and utility rate structure. This method has been extensively followed for more than two decades for evolving and validating energy standards of different constructions in many countries around the globe.

### 15.3.3.5   Ecotect

Ecotect is a study tool that works on the principle of visible predictions and is capable of handling 3D geometric models of any dimension and complexity and is capable of providing feedback on performance analysis at the earliest stages of the building design process. It is equipped with real-time animation features that work on the principle of integrated synergistic auditory and sun beam tracing that updates in actual time with variations to construction engineering and material properties (Reza 2015).

### 15.3.3.6   E-Quest

E-Quest provides high quality results by integrating a building innovation strategy with suitable energy optimization measures and an explicit module for exhibiting the outcomes with an updated and efficient DOE 2.2 building energy use simulation program. DOE 2.2 performs an hourly simulation of the building based on construction, human mass, and ventilation. This analysis tool is efficient in generating multiple simulations for the users permitting the creation and visualization of the possible outcomes with parallel graphics.

### 15.3.3.7 IES Virtual Environments

IES virtual environments are based on design optimization, depending on comfort criteria and energy use. The virtual environment modules mostly include geometry building and improvisation, prediction of charges, and loads with proper thermal and natural ventilation systems. It also considers the optical designs and energy use involved, along with building evacuation strategies.

## 15.4 CONCLUSION

The natural environment's deterioration is inevitably affecting people's health in many parts of the world. Green cleaning is a new, promising approach for environmental cleaning that aims to benefit human health by reducing environmental damage. Green cleaning also aims to improve the cleanliness of the healthcare environment, thereby preventing infection. The components for green cleaning are chosen based on various areas' vulnerability to the risk of cross-transmission of infections. The preferred cleaning products are the ones that use the least amount possible of chemicals. The chapter outlined the various ways of implementing sustainability in healthcare through the different dimensions of sustainability practices involving the environment, customers, employees, and community.

The environmental effects upstream from healthcare delivery are a major area of concern for the stakeholders in the healthcare system. Healthcare services are largely dependent on an appreciable amount of natural resources, including common and rare metals, natural reserves of pharmaceutical precursors, rubber, petroleum-based products, biomass, and water. The most commonly used hospital items, like intravenous pumps, x-ray films, and latex gloves, demand complex manufacturing processes, which have a significant impact on the environment. The environmental costs of the healthcare industry consuming natural resources have not been studied in detail. Thus, the extent of environmental damage caused by these activities is difficult to predict. Future studies should aim to determine and analyze the environmental effects upstream from healthcare delivery to achieve and implement a better, sustainable green healthcare.

## REFERENCES

American Society for Healthcare Engineering. 2020. www.ashe.org.
Arayici, Yusuf, and Ghassan Aouad. 2010. "Building information modelling (BIM) for construction lifecycle management. construction and building: design, materials, and techniques." *Nova Science Publishers* pp. 99–118.
Cooper, J. 2007. "Creating efficiencies with building information modelling." *Health-care Design.*

Di Fabio, A. 2016. "Positive relational management for healthy organizations: psychometric properties of a new scale for prevention for workers." *Frontiers in Psychology* 7:1523. doi: 10.3389/fpsyg.2016.01523.

Garvin, M. L. 1991. *Infectious Waste Management a Practical Guide.* Boca Raton (FL): Lewis Publishers.

Garza, Jennifer L., Jennifer M. Cavallari, Sara Wakai, Paula Schenck, Nancy Simcox, Tim Morse, John D Meyer, and Martin Cherniack. 2015. "Traditionally and environmentally preferable cleaning product exposure and health symptoms in custodians." *American Journal of Industrial Medicine* 58: 988–995.

Global Health and Safety Initiative. 2019. www.ghsi.ca.

Green Guide for Healthcare. N.d. www.gghc.com.

Hattenbach, T. 1998. "'Greening' hospitals: an analysis of pollution prevention in America's top hospitals." *Environmental Working Group/The Tides Center and Health Care Without Harm.*

Health Care Without Harm. N.d. http://noharm.org.

Jameton, Andrew, and Jessica Pierce. 2001. "Environment and health: sustainable health care and emerging ethical responsibilities." *CMAJ* 164(3): 365–369.

Markkanen, Pia Markkanen, Margaret Quinn, Catherine Galligan, and Anila Bello. 2009. "Cleaning in healthcare facilities: reducing human health effects and environmental impacts." 14–25.

Oduyemi, Olufolahan, and Michael Okoroh. 2016. "Building performance modelling for sustainable building design." *International Journal of Sustainable Built Environment* 5: 461–469.

Practice Greenhealth. 2020. http://practicegreenhealth.org/.

Quan, Xiaobo, Anjali Joseph, and Matthew Jelen. 2011. "Green cleaning in healthcare: current practices and questions for future research." *Healthcare Without Harm* 18.

Reza Fallahtafti, M. M. 2015. "Optimisation of building shape and orientation for better energy efficient architecture." *International Journal of Energy Sector Management* 9(4): 593–618.

Sarhan, Saad, Christine Pasquire, Amira Elnokaly, and Stephen Petrove. 2019. "Lean and sustainable construction: a systematic critical review of IGLC Research." *Lean Construction Journal* 1–20.

United State Green Building Council. 2020. www.usgbc.org.

# Section VI

## Green Philosophy

# 16 Green Thinking, Green Acting
## *Redetermining the Self as Ecological*

*Gerard Kuperus*

## CONTENTS

## 16.1 INTRODUCTION

In a time of ever more destructive hurricanes, the rising of sea levels, storms in areas that were calm before, intense monsoons, polar vortexes with extreme and dangerous temperatures, droughts, and unprecedented wildfires in different areas of the globe, we can rightfully speak of "climate chaos." We are living in "the sixth great extinction." The term "Anthropocene" has been proposed to define our current geological era. The details of the changes are debated, including the right terminology, but it is clear that anthropogenic climate change has arrived.

Multiple efforts are currently undertaken to curb greenhouse gases, to protect and restore natural environments, to clean up the oceans, to stop pollution, and so forth. Yet, I want to start with the recognition that those efforts are simply not enough to counter the destruction of the planet that is continuing at a rapid pace. In fact, many of the "green technologies" actually contribute to the problem, since they keep the idea alive that technology will find us a way out of this crisis. While it lies outside of the scope of this chapter to discuss the arguments for this, I strongly believe that we cannot engineer our way out of this crisis. We need to radically change our way of living. I might preach here to the choir in this regard, but the larger question is how others can be convinced to join in a much broader effort to reverse the course we are on. However, my question is not about how we can "save the planet" (or more likely our own species – the planet will probably be fine after we depart), but rather why we should do it in the first place.

Considering the status of our planet, it should be clear that the way we have been living cannot continue. We are using natural resources as if their supply is infinite. We collectively pollute the air, the waterways, and the earth beneath our feet, as if this will not have any consequences for natural environments that sustain us or for our own health. We can and have to change all that. In principle, the actual change is not complicated at all: consume less, manufacture products that are designed to last (as opposed to "made to break"), drive less, and so forth. We can institute policies to encourage this. Yet, governments are unlikely to do this since the number one concern of virtually all governments is the economy, which currently relies on consumption. Likewise, people (who elect government officials) need jobs, which are tied to a strong economy. As we witness with the COVID-19 outbreak, a crisis that requires us to prioritize health destroys the economy and leads to massive unemployment and poverty, especially in countries with poor social security. The situation shows how delicate the system is and how vulnerable we all are. Some of us, myself included, have the choice to isolate and stay at home. It is a challenge, but nothing compared to the challenges many others face as they have no choice but to keep working dangerous jobs. Ironically, they are literally risking their health and lives working jobs that often do not even provide proper healthcare. We could fix this, of course. It seems to be a no brainer to take care of those who provide essential services. Yet, the inability to solve this issue shows how broken our political system is and more importantly, the mindset that keeps supporting this system.

A crisis such as the COVID-19 pandemic provided opportunities to ask very critical questions about our society, such as the lack of proper healthcare for those who provide essential services. The reader might think at this point that the subject has derailed and that I am going off on some kind of tangent, but the larger point here is that the way we treat others is directly related to the environmental crisis. We can see this clearly when we think about consumer goods – "non-essential goods" or "wants" as opposed to "needs." The production of those consumer goods compromises the health of people working in manufacturing businesses, destroys the environments used for resource extraction (destroying the habitat of both humans and non-human animals), and pollutes the atmosphere and the earth. This is the great contradiction we face and that we currently live in: in order to satisfy the wants of some, others have to lose their habitat and risk their health and lives. We are all familiar with the economic reasons to consume and I will not repeat them here. The consensus of our society rests on the belief that either the planet is already healthy enough or that a healthy planet and a strong economy cannot be obtained simultaneously. After all, trees are still growing, birds are flying through the sky, and our backyard is full of life. We do not witness the suffering of human and non-human others, and we believe in the lies that corporations present about their commitment to the environment. Yet, deep down, we know better and I suggest we have to cultivate this dormant wisdom that we cannot maintain the path we are currently on. Our children know! The question we then first need to ask is why the economy rules all our decisions. When we do so, the seemingly radical changes that we have to envision will actually be rather evident.

In this chapter, I argue that we can only make the required changes by first of all changing how we understand ourselves. I propose that we rethink who we are as determined through our environment, consisting of both people (such as friends, family, colleagues, community members), as well as built and natural environments. Such a redetermination of ourselves is not exactly new as we find it in Confucian, Daoist, Zen Buddhist, and many indigenous ways of thinking. With the help of those traditions I will argue for a redetermination of ourselves as ecological selves, in which individualism becomes secondary to community and community includes the natural world.

While it might sound as an interesting philosophical exercise to explore this idea of the self, I argue it is the only way in which we can start to act differently. I am certainly not the first one to suggest that our acts are rooted in our thinking. Most famously, Baird Callicott proposed that "environmental philosophy is environmental activism" (Callicott 1995). It is our worldview that determines how we act in the world. I fully agree with this suggestion and my proposal for a green way of acting is thus rooted in a worldview that places ourselves in a larger context. To be clear, I am not suggesting that we should collectively convert to Confucianism, Daoism, or Zen Buddhism, but instead I argue that these traditions can assist us in rethinking our own tradition(s). As Nietzsche clearly indicated, old values and old habits are hard to beat. Yet, as we see the rise of youth movements fighting for a new future (their future) and we recognize that all values and habits are indeed constructed, then reconfiguring and reconstructing values will be the only viable way forward.

## 16.2 PLACING OURSELVES IN A WORLD FULL OF WISDOM

We know so much. We are constantly collecting data and processing them through ever faster computers. We send rovers to explore distant planets, investigating samples of soil and sending data along with pictures. On our own planet we measure the air, water, and soil, and analyze the levels of pollution. We know, for example, the levels of ozone, particle pollution, carbon monoxide, sulfur dioxide, and nitrogen dioxide in the air and know which levels are safe, risky, or dangerous. We can run models to predict when we run out of fossil fuels and how temperatures on the earth are going to change. As we are connected globally, we share knowledge and information within a matter of seconds. We, indeed, know so much.

Yet, wisdom is harder to come by. The great contradiction and irony are immediately clear: we know that the air is often not healthy, that temperatures are going to change on the planet, that we are going to run out of fossil fuels, and yet we are not wise enough to act on this knowledge. We know that we do not have a long-term solution to our energy needs. We know that we cannot keep on living the way we do indefinitely. Again, we lack the wisdom to change. Likewise, we have forgotten (or in many cases systematically destroyed) indigenous wisdom that tells us how to live responsibly and ethically in our particular place (for example, by being in balance with the ecosystems that sustain us) (Kimmerer 2013). Such knowledge, and the underlying wisdom to "listen to" our natural environments is lost to us "knowers."

One way to think about the problem is that we have separated ourselves systematically from other animals and the natural world. We generally assume that nature is only found outside of cities and towns, outside of human civilization. Within this mindset, culture and nature are typically opposed to one another. Even while the word "Anthropocene" indicates that today we have changed the surface of the earth or its geology to such a degree that the whole planet at this point in time is touched by humans, we still maintain a belief of human and nature as separate. As Gary Snyder points out, we have forgotten "cultures of wilderness" and we should develop what he calls "the practice of the wild" (Snyder 1990). This practice brings together Zen Buddhist and indigenous practices, and challenges the dualism of culture and nature (or the wild). Snyder suggests that starting with our wild bodies, we are nature. Yet, that does not mean that we should just be wild in the sense of being irrational. Quite the opposite: we have to listen to the intelligence of the earth, to a mind that is larger than our own.

Here I will pause, since "a mind that is larger than our own" might quickly lead to creationist interpretations or to some kind of idea that the earth is governed by a god or deity. This is not what Snyder or Zen Buddhist traditions seek to suggest. The intelligence of the earth is the collective of beings ranging from microorganisms to mountains and including humans.

While the intelligence of microorganisms and fungi are largely invisible, we can find more obvious intelligent processes right in front of our eyes if we actually care to look. In the famous *Mountains and Waters Sutra*, the thirteenth-century Japanese Zen Master Dōgen describes the intelligence of mountains and waters which interact with one another and in which we have to find our own place. Waters carve rivers and coastlines in mountains. Mountains are constantly undergoing change with the seasons, eroding and becoming over time, through volcanic and tectonic forces. Nothing is permanent. Snyder describes Dōgen's mountains and waters as "the totality of the process of nature" (Snyder 1990, 102).

Many indigenous philosophies suggest that the world is created through the interaction of animals with the elements. Such cosmologies are important in so far as they inform us about our own place. We only exist because of other animals. While this might not literally be the case (a story such as "Turtle Island" is a story, not a fact), these stories provide insights into the intricate relationships we have with the natural world. They indicate that we are connected and dependent on other animals who lived before us. We would not be here without them. Darwin would not have disagreed.

Thus Darwinism, many indigenous philosophies, and Zen Buddhism all point us to an intelligence that is larger than ours. This is not an intelligence or mind of a god, but the very interaction of the different beings with which we share the earth. All point us, albeit in different ways, to the need to live with those other beings. The basic message is that we cannot separate ourselves.

## 16.3   SEPARATION

I will return to Zen Buddhism in the last part of the chapter, but I will now provide a brief analysis of our current understanding of ourselves as separated from both others and the natural environment. The attempt to separate ourselves from nature has been largely successful, at least in our mind, and it shows in our behavior toward one another and toward the environments and communities that sustain us. As far as nature is concerned, in our everyday living we are separated from it. Perhaps we go camping, but that is typically a vacation or a break from the everyday. The camping trip is an experience of nature, but only in order to return again to our home. Our regular being is away from the insects, dust, and mud that we take for granted in order to be awarded with being outside for a while, but that we refuse to accept in our homes, "away from nature." Needless to say, many people avoid camping for exactly these reasons of "discomfort."

Yet, whether we camp or not, nature still makes it into our homes. Even our homes themselves are made out of natural elements that temporarily have taken the shape of a house after having been trees, rocks, cement, ore, and oil (formed out of plants and animals) before returning in some form to the earth. Our bodies are natural and the food we eat is natural—even after processing it in a variety of ways. Much of our food is some form of manipulated nature, yet it is still nature. We keep rejecting the fact that nature is all around us, in us, that we are part of it, that we are nature. Such rejections are ultimately dangerous, as our failure to mitigate climate change makes clear.

Yet, so many people refuse to accept that climate change is real or is exaggerated. When nature pounces at us, we tend to further strengthen the nature-human division. In the example of COVID-19, people quickly resort to conspiracy theories that keep the distinction alive: nature cannot kill us, but rather it must be a biological weapon created by humans and released either on purpose or accidentally. The theories have one thing in common: a virus that kills us humans, separated from nature, cannot be a product of nature. Its origins have to be human. This is the way for us to find comfort in the midst of a natural pandemic. The idea that a virus can jump from a bat (or a swine or bird) to a human is unacceptable because we have lost the insight that these animals are our kin. Indeed, we are part of a world that includes viruses. All life shares the same building blocks and the very fact that we are so much alike makes it possible for us to be the host of a virus.

Not only do we separate ourselves in our thinking from other species and organisms, but we do the same with individual human beings. In the Western world we are encouraged to think of ourselves as individuals who are in many ways unique and independent. Dependency is seen as a negative characteristic and we are constantly trying to develop ourselves as unique individuals. As Ulrich Beck points out, we are constantly building our biography, creating ourselves as unique individuals (Beck 1992). Our students are advised that if they want to stand out on the job market they have to do something extraordinary, such as double major, get a Master of Arts, do several (unpaid) internships, be involved in student clubs, etc. What we often fail to notice is how even in this individualization students are utterly tied to others: their fellow students, their professors and advisors, their family, and the people in the organizations they take internships at. The very idea to become individuals and to create our unique resumes is in itself affirming that we regard ourselves to be individuals independent of others. The urge to become an individual is a mass demand, part of a herd culture. Nevertheless, whether we want to land a job or a partner, we have to show our unique qualifications. Similar to fashion statements, acquired skillsets are part of a massive trend. This raises the obvious question: can one become oneself by following a mass movement? A student studying Environmental Science can pick

Go Green for Environmental Sustainability

up a set of skills involving data collection, data analysis, the use of a geographic information system, and take two internships, one at an environmental organization, the other at a local government agency. Such a resume may land one a job, indeed, but the fact that I can simply create a generic description of the career path of such a student is an indication of the lack of uniqueness. To be clear, all our students are unique individuals, but our society and educational systems seem to mostly reduce that uniqueness by providing standard molds through which we are invited to individualize ourselves "freely." Indeed, when freedom consists of a list of multiple-choice options without an ability to question those options and questions, we can question the very notion of liberty.

It is certainly the case that all species alter their habitats. The growth of a species is typically limited by the amount of natural resources available or predators. Ants in some cases (often linked to a mutation in its genes) grow beyond the carrying capacity. The human species seems to be the only species that moves natural resources globally. In doing so it destroys the living conditions of other members of its own species. We can find some examples of this in other species and we often refer to this as the survival of the fittest, a term favored by the so-called social Darwinists. Yet, Darwin and many Neo-Darwinists, especially those who studied animals in harsh conditions (such as Peter Kropotkin), emphasize that most animals cooperate. We could say that the very idea of survival of the fittest is often used as an excuse for acting the way we do (De Waal 2010).

What is arguably more stunning is that we are the only species that destroys its own living conditions. Even more alarming, we do so knowingly. We know that over-fishing is a tremendous issue and will only be worse in the future; that our energy consumption through burning fossil fuels is polluting the very air that we breathe; that clear-cutting forests is not only destroying the habitat of members of other species, but also further reduces air quality. Global warming, climate chaos, and exposure to pollution already have imposed tremendous tolls on human lives. We know it will all get worse, yet through the philosophical mindset that regards the successful self as one that separates from others, we have convinced ourselves that we are independent and autonomous.

## 16.4  "NON-WESTERN" IDENTITIES

In the previous section, I have argued that the Western sense of freedom is in practice rather superficial and leads to destructive practices both toward natural environments and others. We can question if within the structures that society provides us, we can be independent in our identities, even while it is constantly emphasized that we are unique, and our independence is celebrated. Other traditions emphasize that all parts of our identity are relational. Confucianism famously emphasizes relationships: I cannot be a son, a father, a husband, a professor, without having parents, children, a partner, students, etc. This is a simple insight, yet within the context of a Western society we have largely forgotten the relational aspects of our being while busy in our attempts to construct our unique individual selves. We are in these attempts so self-centered that we forget that we *are* only in so far as we *are in a world*.

Confucianism is often associated with filial piety (*xiao*) which is indeed a central aspect of its ethics, but it would be a mistake to stop there. Confucius repeatedly points out the relation with others, among others in this famous quote: "When you meet someone better than yourself, turn your thoughts to becoming his equal. When you meet someone not as good as you are, look within and examine your own self" (*Analects IV,* 17). One straightforward reading could relate this back to the Christian saying that one should first remove the beam out of one's own eye before removing the speck out of one's brother's eye. Even if one is better than someone else one should not be satisfied: the flaws in others are likely found in oneself to some degree. What is most of all important here is that we are never alone in shaping who we are and in becoming better human beings.

The underlying idea is that benevolence resides in us already, but it needs to be cultivated. What in Confucianism is known as "the sprouts of the virtues" needs proper nourishment. We all have the virtues already naturally contained in us, but without taking care of or cultivating them they will not flourish. Especially Mencius makes clear that such cultivation, while it cannot be forced, requires the right environment, consisting of good leadership and education.

Confucianism shares important concepts with Daoism, including that of "the Dao" or "way." The way is in a sense already there (just like the sprouts) and we just need to find it. In the *Dao De Ching* this is phrased as follows:

The way of nature is unchanging.
Knowing constancy is insight.
Not knowing constancy leads to disaster.
Knowing constancy, the mind is open.
With an open mind, you will be openhearted.
Being openhearted, you will act royally,
Being royal, you will attain the divine.
Being divine, you will be at one with the Tao.
Being at one with the Tao is eternal.
And though the body dies, the Tao will never
pass away (Lao-Tze 1989, 16)

This, first of all, places us humans within a much larger context. We die, the *Tao* does not. We are impermanent, the way of nature is unchanging. Although the word "divine" is used here, it should not be associated with a God or creator. Daoism does not invoke intelligent design in a Judeo-Christian sense. Instead the *Tao* has to be understood as the natural order of the universe. It is reality and we are temporarily a tiny part of it. The idea that somehow, the *Tao* and the virtues are already there is crucial in thinking about ourselves in relationship to the world. Certainly, the tradition of Western philosophy, which shapes how we understand ourselves, presents some similar ideas. Plato, in the *Meno,* discusses how mathematical knowledge is regarded as inherent and in Aristotle's philosophy virtue is something in need of cultivation, so that we reach a *hexis,* often translated as "disposition" or a second nature (Aristotle 2014). Yet, we find here also a fundamental difference with the Confucian and Daoist traditions, since the emphasis in his analysis of the virtues centers on building a character, a self. This self learns to enact virtues by doing the right thing, at the right time, in the right amount, and for the right reasons. While these reasons are largely about the factors external to oneself, the last one is completely internal and the most difficult factor to assess. For Aristotle, the right reason is to do something because it is good. So, one should not donate money to a hospital so that a wing of the hospital will bear one's name. Instead one donates because the hospital needs this amount of money, they need it now, and the benefactor has it available. Doing good for the right reasons will bring joy, but that in itself is not the reason one should be doing it. So, everything ultimately leads back to the self. It is true that for Aristotle virtue is a political science, in so far as it always involves other people and ideally as many as possible, yet virtue does ultimately lie in the self.

The emphasis on the self relates to other important differences between the Western and the Daoist/Confucian accounts. For Daoism the way—the natural order of the universe—is already there. It is eternal and we humans are encouraged to find it. Instead of turning inside us to develop a character (as in Aristotle) we first of all have to turn to reality. This emphasis on the significance of what is outside of us, has significant implications for our relationship to the external world, including other people and the environment, natural, or built. The word "environment" might in fact be problematic since it provides a separation, as opposed to an essential relation that determines the self.

For Confucianism the virtues are in us and need to be cultivated. This might seem close to Aristotle, but with one significant difference. At the beginning of Book II of the *Nichomachean Ethics* he famously describes a stone that by nature always moves downward. No matter how hard we try we cannot make a stone move upward. Likewise fire always moves up (Aristotle 2014, II, 2). Aristotle's point is that we are not stones or fire, and that unlike inanimate beings, plants, and animals, we can develop dispositions that are different from the nature one is born with. Thus, along these lines, Aristotle argues that "none of the virtues of character arise in us naturally" and more specifically "virtues arise in us neither by nature nor against nature. Rather we are by nature able to acquire them" (ibid). What is interesting here is that while initially it might seem that Confucianism and Aristotle are close and that the difference is subtle, they are in fact essentially different. For Aristotle we move beyond our natural constitution. We develop our second nature of virtuous disposition. Nature provides the possibility to develop it, but we also have to move beyond what is naturally given. For Confucianism, on the other hand, virtues are already in us, given to us naturally. Daoism, as we have seen, tells us to become one with the way of nature, turning to a natural unity.

We can then suggest that Confucianism and Daoism urge us to act in harmony with the natural order that is already given. Aristotle, on the other hand, urges us to create a nature—a second nature—that is an improvement over what has been given. Instead of finding the self in the natural order, Aristotle urges us to develop a self that overcomes nature, a second nature. At least, this is how Aristotle is often interpreted. I will return to this point in the conclusion, where I will provide a slightly different reading.

The notion of the self as discussed above suggests that we are independent. Arguably, such an idea of overcoming nature is found at the foundation of Western thinking. In order to juxtapose this idea I will now further unpack the idea of the dependence of the self through our relation to our natural environment, I will now return to Zen Buddhism. Writing about the self, Dōgen states: "To study the Buddha way is to study the self. To study the self is to forget the self. To forget the self is to be actualized by myriad things"(1985). While it is often assumed that in Buddhism there is no self, we find here a challenge of that idea: one cannot study or forget a self if it does not exist. In fact, Dōgen urges us to study it in order to forget it. It would be a mistake to regard this forgetting as a cancellation or deletion of the self. Forgetting is not the end. We only forget the self as an independent entity that we are so attached to, in order to instead actualize the self through the myriad of things. What forgetting then means in this context, is to radically change our perception of the self.

As we can see, the Confucian, Daoist, and Zen Buddhist traditions share significant ideas. While I am not suggesting they are the same (there are very obvious differences), the three traditions have a lot in common partially because in its move from India to China (and eventually Japan), Buddhism was highly influenced by the Chinese context in which both Daoism and Confucianism shaped the ideas that formed Chan and Zen Buddhism. We can arguably see this influence, among other things, in thinking about a self as constituted through its context, through the myriad things, as Dōgen puts it. Along these lines Buddhism provides a notion of enlightenment that is explicitly not a transcendence to some other realm. Instead, enlightenment is opening one's eyes and mind to what is already there, finding among other things the essential interconnections between self and other. This notion of enlightenment is compatible with the Daoist conception of the way, which is eternal and already there. We just have to learn to see it.

In the earlier mentioned "environmental" fascicle entitled "The Mountains and Waters Sutra," Dōgen provides all of the insights I have outlined above. First, the sutra is not a human discourse about mountains and waters. Instead, it is the sutra of the mountains and waters – the natural order of the universe, if you will. Dōgen tells us here that if we listen carefully to the mountains and waters, this is what they tell us. This is their story. What, then, is their story? First, they are not set up in some dualistic way, with a mountain here and a river over there. Hydrologists know well that a river does not stop at the bank. The majority of the river flows often under the bank, through the gravel, rocks, sand, and soil on which we might be walking or even build a house (or a monastery). Yet, Dōgen takes his insights further, suggesting that the opposition between flowing water and unmovable mountains is a false one. He challenges this opposition through the image of walking mountains: "The blue mountains are constantly walking" (Dōgen 1985, 24). We can agree that mountains are undergoing change, on a geological timescale, but also through the seasons, even on a day to day or even an hourly basis. The weather can change quickly in the mountains. In relation to water, a fluid entity, we find something interesting: water is constant, strong, and even stronger than rock. It carves its way into mountains and erodes coastlines.

Dōgen, thus, challenges our common perceptions of mountains and rivers, finding the one in the other. Why, though, would he do so by using the human quality of walking? He even tells us to study this carefully, and if you disagree "you do not know your own walking" (Dōgen 1985, 24). This last addition provides a key insight, pointing to the unity between the human and the non-human world. We are the mountains and waters, and the mountains and waters are who we are. We are all made out of the same material and we are a part of the ecosystems they provide. It should be clear at this point that Dōgen does not only refer to literal mountains, but to the natural world of which humans are a part. As mentioned above, Snyder accurately calls the mountains and waters "the processes of the earth" (Snyder 1990, 102).

What now does this mean for the self? As I have argued elsewhere, Dōgen provides us an idea of a self as "contextualized" (Kuperus 2019). That is, he does not suggest that there is no self, but rather that the self is, as we saw in the earlier quote from *Genjo Koan*, "actualized through the myriad things." I cannot be without a context that determines me. That context is partially filial and social (as particularly emphasized by Confucianism) and it is partially natural (as emphasized here by Dōgen). Specifically, through Daoism and Zen Buddhism one can suggest that "community" is not limited to the human world. We are in a community with the natural world. Even more, our human community is part of this larger community. And we as individuals are part of both. The self is not first of all an independent, autonomous being. It can only develop (or actualize) some form of independence within a network of relationships it depends on in terms of its identity and essence.

## 16.5 CONCLUSION

I have argued in this chapter that within a Western context we tend to individualize and consider the ultimate goal to be independent and autonomous. Our destructive tendencies toward our natural environments can be seen as a symptom of the urge to be independent. Confucianism, Daoism, and Zen Buddhism show in complementary ways how such independence and autonomy is delusional. We can only actualize ourselves as individuals within the context of our social and natural environments. This implies that we can never be truly independent. Even more, to think that we are independent and to forget the simple truth that we are constituted by our context, leads us to practices that are ultimately self-destructive. We need our human and natural communities to thrive.

Does this mean that we should all become Confucianist, Daoist, or Buddhist? Besides the fact that such a massive conversion would be straight-out impossible, we can find many examples of environmental destruction in regions that are defined through Confucianism, Daoism, and/or Buddhism. Likewise, one could object that Western approaches do not by definition have to be destructive. We find environmentally sustainable approaches in Christian traditions. Pope Francis' *Laudato Si* is an excellent example of this. He calls for an integral ecology, in which social and natural ecologies are regarded in a unity. We need exactly these kinds of conversions within our own systems of belief (whether they are scientific, religious, and/or cultural). Such conversions can happen with the guidance of other traditions. Often this means, as Pope Francis shows, that a less self-centered and much more integral or contextualized understanding of ourselves already exists within our own tradition.

Thus, going back to Aristotle and Plato one could argue that those aspects that emphasize the independence of the self have been overemphasized, while the communal and environmental aspects have been disregarded. In the *Crito*, Plato makes it very clear that he would not even exist without the city of Athens, and that he has to be grateful for everything this city gave to him. Aristotle regards ethics as a political science and was extremely interested in the natural world, in which he did find souls not only in humans and animals, but also in plants. All life, he suggested, functions on shared principles, such as growth and nutrition. Such a fundamental "kinship" could potentially be the basis for a new understanding of ourselves.

I thus suggest that the problem is not the tradition itself, but how it has been interpreted. What is emphasized (as opposed to what is left out) then constitutes our worldview. To be unique and independent as an individual or the idea of the self-made individual are some of the ideals that have led us in this direction. Lack of engagement in our communities, greed, and selfishness, are some of the ways in which this idea of independence manifests itself. Whole communities are left to waste because we somehow do not acknowledge them, even while the people living in those communities provide all kinds of essential services. The same is true for our natural environments, which we also regard as something separate from our existence, our well-being, and our self.

In order to first of all recognize the natural communities of forests, clean air, soil, water, and biodiversity as essential to both our individual and collective thriving, we have to become community members. These communities exist of both human and non-human others, of non-human animals, plants and trees, constituting ecosystems of which we are a part. My suggestion in this chapter is that a non-self-centered and non-dualistic approach might help us to think differently about the relationship to our natural and human environments. In fact, the suggestion is to step beyond a separation of nature and human, or nature and culture. Different traditions can provide us steps to overcome this separation. While other traditions can provide us important ways to start this process, ultimately, we have to find ways to rethink ourselves within our own traditions. In this chapter, I have shown how such an engagement can be undertaken in which we take a step away from our own tradition, in order to critically reassess it.

## REFERENCES

Aristotle. *Nichomachean Ethics*. 2014. Indianapolis: Hackett.
Beck, Ulrich. *Risk Society: Towards a New Modernity*. 1992. London: Sage Publications.
Callicott, Baird. 1995. "Environmental Philosophy Is Environmental Activism: The Most Radical and Effective Kind." In *Environmental Philosophy and Environmental Activism*, ed. D. E. Marietta, and L. E. Embre, 19–36. Rowman & Littlefield.
de Waal, F. 2010. *The Age of Empathy, Nature's Lessons for a Kinder Society*. New York: Three Rivers Press.
Dōgen. 1985; "Genjo Koan." In *Moon in a Dewdrop*. San Francisco: North Point Press.
Kimmerer, R. 2013. *Braiding Sweetgrass, Indigenous Wisdom, Scientific Knowledge and the Teachings of Plants*. Minneapolis: Milkweed Editions.
Kuperus, G. 2019. "A Contextualized Self: Re-Placing Ourselves Through Dōgen and Spinoza." *Comparative and Continental Philosophy*. Volume 11.3.
Lao-Tze. 1989. *Tao Te Ching*. Feng, Gia-Fu & Jane English (tr.). New York: Vintage Books.
Snyder, G. 1990. *The Practice of the Wild*. San Francisco: North Point Press.

# Index

## A

Adsorption 32, 34, 35
Agricultural nanotechnology 58, 59
Anthraquinone dye 4–6, 10–12
Anthropocentric 205, 219, 220
Artificial Intelligence (AI) 102–107, 112–114
Azo dyes 31, 32, 38, 41
*Azotobacter* 18, 20, 23

## B

BILLDESK, 82
Biofertilizers 18, 19, 22, 23
Biologically derived pesticides 52–54
Biopesticides 18, 23, 52–54
Bioremediation 4–9, 11, 33–37, 39–42
Bio-stimulants 15–18, 20, 23
Blockchain 102, 104
Building information modeling (BIM) system
        135, 233

## C

Carbon
        pricing 70, 72–74, 76
        taxes 72
Carbon-neutral 138
Carbon usage effectiveness (CUE) 93, 94
Carpooling 117–130
Circular electrical and electronic equipment
        (CEEE) 134, 142
Climate
        change 69–73, 77
        risks 69–72, 74
Cloud computing 88, 90
Combined heat and power systems 138
Concrete composition 167
Confucianism 241, 244–248
Consortium 33, 36, 39–41
Consumer sentiment data 152
CooA 7, 9, 12
Corporate Environmental Reporting Index
        (CERI) 178, 180
Cost comparison 169
CSR 85

## D

Daoism 241, 244–248
Data center infrastructure efficiency (DCIE) 93, 94

Data center performance efficiency (DCPE)
        93, 94
Dehaloperoxidases (DHPs) 4–12
Digital India 83
Drones 102, 106, 109, 110, 112
Dye-decolorizing peroxidases (DYPs)
        5, 6, 8–12

## E

Effluents 31–33, 36–41
Electronic waste (e-waste) 92, 95
Embargoes 154
Energy modeling 234
Environment and Economics Group 135
Environmental impacts of agriculture 59
Environmentally friendly projects 85
Environmental
        concern 202, 204, 207, 220
        impact 203
        knowledge 204, 206, 219
        management 182, 186–192
        policy 182, 186–191
        reporting 175–180, 191, 192
        sustainability 118
Extremophiles 37–41

## F

FixL 9, 10, 12
Forest Fire Danger Index (FFDI) 150

## G

G4 Guidelines 179, 180, 198
Geographical information system (GIS) 103
Generation 2030 134
Genetically modified (GM) crops 54, 55
GGBFS 164
Green
        building 233
        chemistry principles 50, 51
        cleaning 227
        computing 87
        data center 88, 93
        health 226
        marketing 202, 203
        technology 102
Global positioning systems (GPS) 103
Globin couples sensors (GCSs)
        7, 10, 11, 12